Environmental Science & Engineering Laboratory Manual

Second Edition

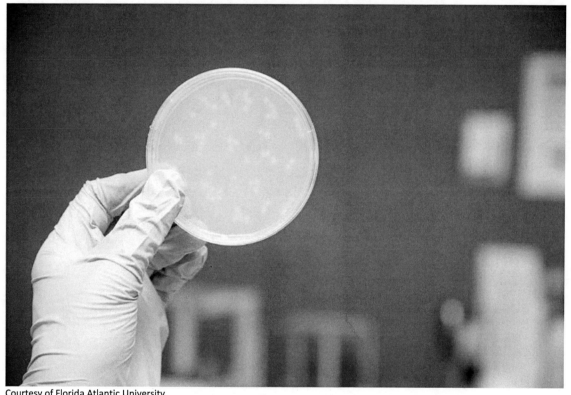

Courtesy of Florida Atlantic University

By
Daniel E. Meeroff, Ph.D.

www.kendallhunt.com
Send all inquiries to:
4050 Westmark Drive
Dubuque, IA 52004-1840

About the Author

Courtesy of Florida Atlantic University

Daniel E. Meeroff, Ph.D., is a professor in the Department of Civil, Environmental & Geomatics Engineering at Florida Atlantic University (FAU). His area of specialization is environmental engineering, specifically, water and wastewater engineering, water chemistry, environmental microbiology, solid/hazardous waste management, sustainability, and pollution prevention. Dr. Meeroff is the founder and director of the Laboratories for Engineered Environmental Solutions (Lab.EES) at FAU (http:\\labees.civil.fau.edu). He earned his bachelor's degree in environmental science from Florida Tech and his master's and Ph.D. degrees in civil/environmental engineering from the University of Miami. Dr. Meeroff designed the Environmental Chemistry laboratory classroom and all of the experiments detailed in this manual. Dr. Meeroff was awarded the Teaching Prize for Excellence in Undergraduate Education (2017) for the second time (2011), the Distinguished Research Mentor of the Year (2015), the John J. Guarrera Engineering Educator of the Year for North America in 2014, and the FAU Distinguished Teacher of the Year in 2014 (as voted by the student body), which is the highest teaching honor at the university.

Table of Contents

List of Figures

List of Tables

Preface

This laboratory manual provides practical hands-on experience designed to help students understand the theory and practice of environmental engineering principles. It places emphasis on material tested as part of the Fundamentals of Engineering Exam (FE). Civil and environmental engineers rely on laboratory results to make decisions about design and comply with regulations. Experiments like those found in this manual provide valuable insight into how laboratory results are obtained, used, and reported in the professional world. The manual has been reviewed multiple times for clarity of language and pedagogical effectiveness by past students from Florida Atlantic University to ensure a successful laboratory experience.

Laboratory Safety

© Shutterstock.com

Safety must be placed at the highest priority level during all laboratory sessions.

Everyone is responsible for laboratory safety and a safe laboratory environment for all students, teaching assistants, and instructors.

All persons working in the laboratory must receive training to become knowledgeable about potential hazards in the laboratory and become familiar with the requirements of the Occupational Safety and Health Administration's (OSHA) Laboratory Safety Standard as specified in 29 CFR§1910.1450, Occupational Exposure to Hazardous Chemicals in Laboratories.

Important issues for laboratory safety include the following topics:

- Regulations
- Standard Operating Procedures (SOPs) and Control Measures
- Personal Protective Equipment (PPE)
- Chemical Storage and Hazard Identification
- Material Safety Data Sheets (MSDSs)

The most important source of information regarding laboratory safety at your institution is the Environmental Health and Safety Office, which provides training, helps identify hazards, and maintains current safety data sheets on chemical hazards.

Find out the contact information for the Environmental Health and Safety Officer for your institution and write it here:

_____

_____

_____

_____

_____

_____

Although the Environmental Health and Safety Office is an excellent resource, you are ultimately responsible for understanding the proper procedures to be followed to limit exposures and to protect yourself from chemical hazards, including the proper precautions and protective equipment that must be used.

© Shutterstock.com

The Role of the Laboratory

The laboratory's primary role is to provide **qualitative** and **quantitative** data used in decision making or engineering design. Laboratory data are critical in engineering decisions on process modifications and regulatory compliance. Data collected must be accurately described and representative to be of value. Extreme care must be taken because decisions that impact human and environmental health, safety, and welfare will be based on laboratory findings, such as those you will conduct as part of this course. Thus, learning to perform laboratory tests on water, wastewater, air, and solid wastes plays an important role in the engineering profession.

Laboratory equipment is expensive, and great care should be taken when using any instrument, glassware, or chemical reagent. Handle and store laboratory glassware to avoid damage, and never use broken glassware. If your equipment appears damaged, broken, or malfunctioning (even if it was your fault), immediately let your supervisor know.

Decontaminate and properly dispose of damaged/unwanted glassware according to the standard operating procedure based on the specific chemical, biological or radiological hazards that may be present. Wash all glassware after each experiment with laboratory-grade detergent and rinse with tap water. A final rinse should be done with deionized or distilled water.

All equipment must be properly maintained to provide reliable experimental results. Make certain that equipment is clean before and after use. So always maintain your equipment as if it was your own. Use equipment only for its designed purpose.

© Shutterstock.com

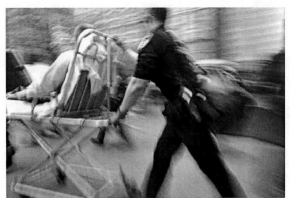

© Shutterstock.com

Maintain a general sense of safety in your work zone to prevent accidents, injuries, and illnesses. If a serious accident or emergency occurs, keep your composure, remain calm, and follow these important steps:

- **Call 911** for immediate medical assistance or any emergency.
- **Give your precise location**.
- Explain the type of injury or incident.
- Notify the supervisors in charge (faculty, teaching assistants, staff, etc.).
- Notify the Environmental Health and Safety Office
- Initiate life-saving measures if required (and you are properly trained to do so).
- Never move an injured person unless there is danger of further harm.

It is critical to know the exact location of the laboratory, so if an incident does occur, you can give the first responders accurate information to arrive to the scene on time.

Write here the exact location of the laboratory:

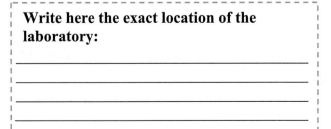

_____
_____
_____
_____

© Shutterstock.com

Be aware of the potential dangers to your health and the well-being of others while in the laboratory. Some of these hazards include:

- Inhalation of fumes, mists, or gases
- Skin absorption of chemicals
- Eye injuries
- Chemical burns
- Slips and falls
- Fires and electrical shock

If exposed to a laboratory hazard, follow these important, potentially life-saving steps:

- Read the label or MSDS for instructions.
- Contact medical personnel or poison control.
- Do not induce vomiting.
- Get to fresh air if you feel dizzy, nauseous, or burning in the lungs/throat.

© Shutterstock.com

If the case of a spill of chemical or biological reagents, follow these important, potentially life-saving steps:

1. Shut down all electrical equipment.
2. Stop leaks or spills from getting worse.
3. Clean up small spills immediately.
4. Allow trained personnel to handle larger spills.

Basic Safety Rules

Upon entering the laboratory, place all your personal belongings in designated areas, not on the bench tops and not where another person can trip over them.

Some basic "*good housekeeping practices*" will always apply to any laboratory activity:

- **Wash hands before and after laboratory activities**

© Shutterstock.com

- **No eating or horseplay**
  - Food and beverages are prohibited in the laboratory. Do not store food or beverages or used glassware in laboratory refrigerators.
  - Do not eat, drink, chew gum, use tobacco products (smoke, chew, dip), or apply cosmetics.
- **Return all chemicals and supplies to the proper location after use**
  - Chemicals and equipment must be properly labeled and stored.
  - Never pipet directly from a reagent bottle, instead pour out what you need into a clean beaker and pipet from there.
  - Never pour a chemical back into a reagent bottle.
- **Keep the work area clean and uncluttered**
  - Clean up your area before you finish.
  - Wipe down the scale every time after use with a damp paper towel.
  - No experiment is complete until the laboratory has been cleaned, and avoid situations like Figure 1.

Figure 1. Do not just store your dirty glassware in the sink
Source: Daniel Meeroff

If you have to leave the room while the experiment is in progress, remember to:

- Always leave the lights on

- Place a sign on the door or at your work station, with the name and contact information of the principal investigator, along with any notes such as *"Do not touch."*
- Provide for containment of toxic substances in the event of failure of a utility service to prevent accidents, fires, explosions, or other hazards.
- Avoid working alone. Prior approval in writing from your supervisor is required before working alone in a laboratory.

Dress Code

If you are not wearing the appropriate attire, you will not be allowed to participate in the laboratory and will receive a zero grade for the session. All students and personnel should be dressed according to these rules:

- Appropriate long pants so that legs are covered to the ankles.
- Closed-toed shoes and socks must be worn.
- Long hair, loose clothing, and dangling jewelry must be confined.
- Cosmetics and other personal care products may react with chemicals used in the laboratory. Avoid the use of these products on laboratory days.
- Contact lenses should not be used in the laboratory setting. Fumes may penetrate the lens and become lodged between the eye and the lens, causing severe damage. It is recommended to use glasses instead on laboratory days.
- Open-toed shoes, high-heel shoes, short pants, and skirts are strictly prohibited to wear during laboratory activities.

Personal Protective Equipment (PPE). Adhere strictly to the rules regarding **personal protective equipment**

(PPE). It is your responsibility to wear the proper gloves and either a lab coat or a lab apron, as required. If you need eye protection, gloves, face shield, air quality protection, or other PPE you should not assume it will magically be provided for you. Instead, you should be prepared with what you need before you come into the lab.

- **Eye Protection.** Use safety glasses with side shield, full face shield, or goggles. Appropriate eye protection must be worn in the presence of the following hazards:

  - Liquid chemicals
  - Acids or caustic liquids
  - Chemical gases or vapors
  - Hazardous light radiation
  - Flying particles, dusts, aerosols
  - Molten metal

Do not wear your goggles around your neck or on the top of your head. Sometimes, the goggles will fog up, but under no circumstances should you remove any parts because this will eliminate the protection.

- **Gloves.** Use the proper type of glove for the hazard or chemical involved. Gloves must be worn in the presence of the following hazards:

  - Hazardous materials
  - Chemicals of unknown toxicity
  - Corrosive materials
  - Rough or sharp objects
  - Very hot/cold materials

© Shutterstock.com

Consult a glove selection chart to identify the appropriate type to use for the hazards that you will be exposed to. If you have an allergy to latex or nitrile, make sure you notify your supervisor right away.

Always inspect gloves for tears prior to each use. Wash gloves before removing. Remove gloves before touching doorknobs, telephones, pens, and other items and before leaving the laboratory. Replace gloves periodically. Reuse of gloves is not recommended, and used gloves should not be left at the workstation.

- **Body Protection**. Protect your feet and legs from corrosive chemicals by following the proper dress code (long pants, close-toed shoes, and using chemical-resistant aprons or coats). Depending on the circumstances, apparel giving additional protection may be required because lab coats only meet the minimum requirements. Note that protective gear must be removed when leaving the laboratory.

- **Fume Hoods**. Use a fume hood if noxious gases are involved (that is, you can smell it). Make sure hoods are operating and that they have been inspected. Do not use the hood unless you can verify that it is working properly with a tissue test. Work with the sash height at the lowest level, and do not block the airflow. Many chemicals are poisonous, and if you can smell them in the air, you may already have received a toxic dose. Never use fume hoods for routine storage, and make every effort not to overload them with containers, equipment, and so forth. It is good procedure to check the MSDS

to determine if a fume hood is necessary. Even if using a hood, you should still use PPE as indicated in the MSDS. If a respirator for an individual is called for, then make sure it is properly fitted first. Acids should be diluted only in the fume hood, and when doing so, always add acid to water; never add water to acid.

Can you spot all of the incorrect usage of PPE in Figure 2?

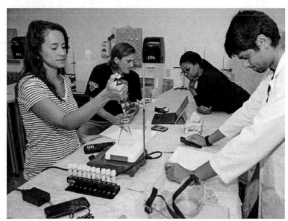

Figure 2. Can you spot all of the incorrect usage of PPE in this picture?
Courtesy of Florida Atlantic University

Safety Showers & Eye Wash Stations. It is critical that you are aware of the location of the eye wash station, safety shower, first aid kit, spill kits, and fire exits. Always make sure that the paths to these areas are kept unobstructed and free from clutter. It should take less than 10 seconds to reach them from any place in the lab.

Material Safety Data Sheet (MSDS). The material safety data sheet (MSDS) contains life-saving information, such as the chemical formula, CAS number, manufacturer's information, hazard warnings, and potential physical/health hazards. It also discusses hazard prevention, accident prevention, exposure control measures, PPE, proper spill response, clean-up measures, first aid, and what to do in case of fire or explosions. In addition, it should also detail the ecological hazards, disposal considerations, regulatory requirements, and toxicological data.

© Shutterstock.com

Always consult the MSDS to ensure that you have chosen the appropriate PPE for the chemical or biological materials that you are working with. Do not assume the MSDS sheet will be available when you are in an emergency situation. Read it beforehand.

Sample Storage/Handling.

Follow these simple rules for handling samples and laboratory materials:

- Bottles and containers must be labeled properly. Each label must have the name of the investigator, name of the chemical, and the date. Label all containers **before** placing solutions in them. Use labeling

tape, and do not write directly on the container.
- Never pipet directly from a primary solution container. Place an aliquot (small portion) of the solution into a small labeled container and pipette from there. This procedure will help prevent contamination.
- Never pour a chemical back into a reagent bottle.
- When not in use, all chemicals must be segregated properly in their designated storage area (e.g., organic, inorganic, acid, and solvent) stored in appropriately compatible containers.
- Fume hoods should never be used for storage of chemicals, including wastes.

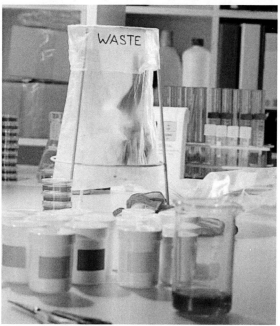

© Shutterstock.com

- Dispose of chemical waste properly. Consult the MSDS to follow the manufacturers' recommendations, and contact your environmental health and safety office for guidance on proper storage and disposal of wastes.
- It is illegal to dispose of hazardous chemicals into a sink unless the

concentration meets the discharge standard. Diluting the waste in order to meet the standard is not only prohibited, it is criminal according to federal law (40CFR268.3) without a valid permit from the US Environmental Protection Agency (USEPA). In addition, evaporating hazardous chemicals in the fume hood is similarly inappropriate.

- Take care to minimize waste because disposal is not free. You must be responsible for managing your waste.

- Place a proper hazardous waste label on the container as soon as the material is identified as waste. The label must show the name of the waste chemical(s), concentration, and the supervisor's name or initials. Unknown or mixed wastes cost more to dispose of. When finished, place the properly labeled waste container in the designated **satellite waste storage area** (Figure 3).

Figure 3. Satellite waste storage area ready for pickup
Source: Daniel Meeroff

- When the waste bottle is full, contact your lab technician or supervisor so they can arrange for pickup.

For specific types of wastes, use the following guidelines:

- **Broken glass**: Discard broken glassware in the designated waste glass container (Figure 4). Make sure to sweep up the broken pieces from the floor, benchtop, or other surfaces.

Figure 4. Broken glass disposal container
Courtesy of Florida Atlantic University

- **Sharps** (e.g., syringes, razor blades, etc.): These items must be disposed of in a designated sharps container. Never use the designated waste glass container or sharps container as a trash can.

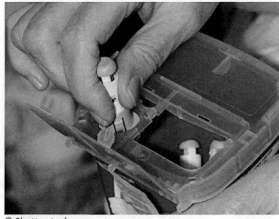

© Shutterstock.com

- **Microbiological or biohazardous waste**: Sterilization (autoclaving or chlorinating) of incubated plates must be performed prior to disposal of biohazardous waste. Decontaminated materials are stored temporarily in the red bag or a designated biowaste collection container. Contact the environmental health and safety office to remove red bag waste from the biowaste collection box. Never use a designated biowaste collection box as a trash can.

Take a moment now to locate the laboratory safety equipment:

Eye Wash: _____

Safety Shower: _____

First Aid Kit: _____

Spill Kit: _____

Fire Exits: _____

MSDS Folder: _____

Satellite Waste Storage Area: _____

Broken Glassware Container: _____

Sharps Container: _____

Biohazard Container: _____

Fire Extinguisher: _____

Electrical Safety

© Shutterstock.com

Ensure proper electrical supply for the intended use, and report any broken/malfunctioning equipment or outlets. Keep liquids away from electrical sources. Do not overload circuits, and avoid fire hazards whenever possible. Never unplug laboratory equipment so you can charge your personal electronic devices.

Fire Safety

Courtesy of Florida Atlantic University

Never leave open flames unattended, and always keep combustible materials separated from heat sources. Take note of the location of fire extinguishers. Never operate a fire extinguisher unless you have been properly trained.

If a fire occurs, use the C.A.R.E. procedure:

- Close doors
- Alert others; activate alarm
- Report fires (call 911)
- Evacuate the building

Report fires immediately. Respond to all fire alarms. Make it a point to know your exit route, and keep it clear of obstructions. Do not use the elevators, stay clear of the building after exiting, and do not re-enter until authorized. Remember, fight fires only if you have been properly trained; otherwise, leave this to the professionals.

Laboratory Safety Quiz

Instructions: Read and answer all questions to the best of your ability.

1. According to the dress code, what clothing should you wear in the laboratory?

2. What type of PPE should you wear in the laboratory?

3. In what cases may you be asked to leave the laboratory?

4. Why are you not allowed to eat, drink, or chew gum in the laboratory?

5. What do you do if you break a piece of glassware?

6. What do you do if a chemical splashes on your face?

7. What do you do if the fire alarm rings?

8. What do you do if your clothing catches on fire?

9. What do you do if acid is spilled on your pants?

10. What do you do if your sample spills on the counter?

Safety Agreement

Instructions: Read the following two-page document, sign both sides, and submit to your instructor before participating in any laboratory session.

By signing this agreement, I hereby declare that I understand and agree to follow all laboratory safety rules established by my institution.

I have attended the laboratory safety lecture provided by my instructor, and I understand that the safety rules included in my laboratory manual are for my information.

I understand that it is my responsibility to make myself familiar with all safety regulations and to follow them at any given time when I am in the laboratory setting.

I understand that I will have to successfully pass the Laboratory Safety Quiz in order to be allowed in the laboratory.

I understand that in any accident that is caused by my disobedience of any laboratory safety rule, I am solely responsible for all consequences.

_____   _____
Signature of Student   Date

_____   _____
Printed Name of Student   Student Number

Safety Agreement

By signing this agreement, I hereby declare that I understand that it is my full responsibility to do the following:

Attend all laboratory sessions for which I am registered for, or I will receive a zero grade for the laboratory.

Keep track of the laboratory assignments and complete them according to the guidelines established in the laboratory manual. Deadlines are to be met or I will accept the consequences. No late papers will be accepted.

Make myself familiar with the experiments and come fully prepared to all laboratory sessions.

I also understand that it is my right to:

Visit teaching assistants and lab instructors during their office hours to seek additional guidance.

Discuss my grades with the instructor at any time throughout the course.

_____   _____

Signature of Student   Date

_____   _____

Printed Name of Student   Student Number

Source: Daniel Meeroff

Make a habit of taking proper notes and recording data in the proper tabular form while conducting the experiment. You are expected to describe your work carefully enough so that it can be reproduced and checked by others. The best proof for a new scientific result is that it can be reproduced.

A laboratory notebook is the instrument to make certain that new discoveries are well documented. Please keep your own, separate notebooks to record your data, and record all information using a waterproof pen. Use a title at the beginning of each page and put the date and your initials on each page.

Warning! Surprise notebook checks will be conducted from time to time.

© Shutterstock.com

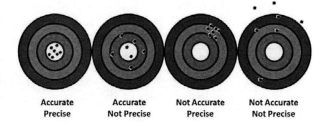

| Accurate | Accurate | Not Accurate | Not Accurate |
| Precise | Not Precise | Precise | Not Precise |

Source: Daniel Meeroff

An understanding of statistics is essential for determining the accuracy and precision of quantitative analytical data. **Accuracy** refers to the *correctness* of a given analysis. **Precision** refers to the *reproducibility* of an analytical procedure. A measure of accuracy can be obtained by analyzing a sample of known concentration (standard) and noting its deviation from the known value. A measure of precision is obtained by repeating the sample analysis multiple times (replicates) and noting the standard deviation from the mean. Some additional statistical terminology is defined as follows:

- **Mean** ($\bar{x}$). The technique of "taking an average" by adding the numerical values (x_1, x_2, x_3, … etc.) of an analysis and dividing this sum by the total number (n) of measurements used.

$$\bar{x} = \frac{\sum_{i=1}^{n} x_i}{n}$$

- **Percent Error**. The percentage that the measured value deviates from the expected or known value. This is determined by taking the difference between the known and measured value and then dividing by the known value, expressed as a percentage.

$$Percent\ Error = \frac{|(true\ value - measured\ value)|}{true\ value} \times 100\%$$

If the true value is not known, we need a different approach. For example, if we have two values obtained using two different methods, then the error is determined by taking the difference between the first measured value and the second measured value and then dividing by the average of the measured values.

$$Percent\ Error = \frac{\left|(measured\ value_1 - measured\ value_2)\right|}{average\ measured\ value} \times 100\%$$

- **Standard Deviation (s_x).** The measure of the distribution of values about the mean. The standard deviation for small data sets ($n < 20$) that are normally distributed can be calculated in five steps:

  1. Determine the mean ($\bar{x}$).
  2. Subtract the mean from each measured value ($x_i - \bar{x}$).
  3. Square each difference $(x_i - \bar{x})^2$.
  4. Sum the squared terms in step 3 and divide by ($n - 1$).
  5. Calculate the square root of the average found in step 4.

$$s_x = \sqrt{\frac{\sum_{i=1}^{n}(x_i - \bar{x})^2}{(n-1)}}$$

- **Confidence Limits.** The interval around the mean within which the true result can be expected to lie with a stated probability. Confidence limits for small data sets are estimated by multiplying the standard deviation times t and dividing by the square root of n.

$$Confidence\ Interval = \bar{x} \pm \frac{ts_x}{\sqrt{n}}$$

The value of t is determined from Table 1.

Table 1. Values of t for Various Levels of Probability (Skoog et al. 2013)

Degrees of Freedom ($n-1$)	Factor for Confidence Interval, %				
	80	90	95	99	99.9
1	3.08	6.31	12.7	63.7	637
2	1.89	2.92	4.30	9.92	31.6
3	1.64	2.35	3.18	5.84	12.9
4	1.53	2.13	2.78	4.60	8.60
5	1.48	2.02	2.57	4.03	6.86
6	1.44	1.94	2.45	3.71	5.96
7	1.42	1.90	2.36	3.50	5.40
8	1.40	1.86	2.31	3.36	5.04
9	1.38	1.83	2.26	3.25	4.78
10	1.37	1.81	2.23	3.17	4.59
11	1.36	1.80	2.20	3.11	4.44
12	1.36	1.78	2.18	3.06	4.32
13	1.35	1.77	2.16	3.01	4.22
14	1.34	1.76	2.14	2.98	4.14
**	1.29	1.64	1.96	2.58	3.29

So let's do an example.

Given: The following masses:

- 36.78 mg
- 36.80 mg
- 36.87 mg
- 36.94 mg

Find:
Calculate the mean, standard deviation, and 95% confidence limits.

Expected:
$\bar{x} = 36.80$, $s_x = 0.1000$, $C.I. = 36.70 - 36.90$

Solution:
Mean:

$$\bar{x} = \frac{\sum_{i=1}^{4}(36.78 + 36.80 + 36.87 + 36.94)}{4} = 36.85\ mg$$

$$\bar{x} = 36.85\ mg$$

Standard deviation:

$$\sum_{i=1}^{4}\left[\begin{array}{l} (36.78-36.85)^2 + (36.8-36.85)^2 \\ + (36.87-36.85)^2 + (36.94-36.85)^2 \end{array} \right] = 0.0159$$

$$s_x = \sqrt{\frac{0.0159}{(4-1)}} = 0.0700 \, mg$$

Confidence interval:

From Table 1, $t = 3.18$ (degrees of freedom = 3 @ 95% confidence interval)

$$Confidence \ Interval = \overline{x} \pm \frac{ts_x}{\sqrt{n}}$$

$$= 36.85 \pm \frac{(3.18 \times 0.0700)}{\sqrt{4}}$$

$$= 36.85 \pm 0.11 \, mg$$

Curve Fitting

© Shutterstock.com

Data can be plotted on a graph, but not all data fall on a straight line. We can draw a straight line through as many data points as possible using a straight edge, but a better approach involves calculating the line of **best possible fit** through the data.

A straight line relationship is given by:

$$y = mx + b$$

- y represents the dependent variable (e.g., concentration).

- x represents the independent variable (e.g., absorbance, NTU, peak area, time, etc.).
- m is the slope of the curve.
- b represents the y-intercept.

Mathematically, the best-fit line through a series of data points is the line for which the sum of the squares of the deviations of the data points from the line is minimized. This is known as the **method of least squares**. Computation of the best-fit line by the method of least squares can be accomplished using Microsoft© Excel

Here is how you do this:

1. Plot the data in Microsoft© Excel in side-by-side columns.
2. Highlight your data and select the appropriate chart and formats by going to Insert>Chart.
3. Once the data is plotted to your satisfaction, left-click one of the data points in the chart.
4. Now right-click and select *Add Trendline*, and select Linear on the Type tab.
5. Select *Display equation on chart* **AND** *Display R-squared value on chart*.

The measure of how well the data fit to a straight line is given by the **goodness of fit parameter** (r^2). This value is computed by subtracting, from unity, the ratio of the variance of the y_i data points with respect to the fitted line (y) over the variance of the y_i data with respect to the y-average value.

If the r^2 value is close to unity (1.0), this signals that the data fit is linear. The closer the value gets to zero, the worse the fit. The example shown in Figure 5 illustrates an excellent linear correlation.

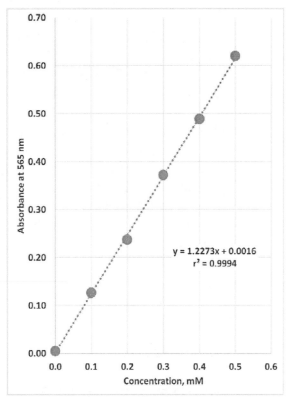

Figure 5. Example of a linear correlation with excellent goodness of fit
Source: Daniel Meeroff

The graph shows: $y = 1.2273x + 0.0016$, $r^2 = 0.9994$, with Absorbance at 565 nm on the y-axis and Concentration, mM on the x-axis.

Significant Digits

© Shutterstock.com

The term *significant digits* refers to the precision or certainty of a value. It is incorrect to report a result with too many significant digits. Here are some helpful rules:

- The number of significant digits is always equal to or greater than the number of non-zero digits.
- A zero counts when:
  - Used between other non-zero figures (Ex. 4006)
  - Used after (to the right of) other non-zero figures in the decimal point (Ex. 3.00)
- A zero does not count when:
  - Used just to locate the decimal point (Ex. 0.00001)

These conventions are best explained by using examples (see Table 2).

Table 2. Useful Examples of How to Properly Understand Significant Digits

Number	Significant Digits	Notes
12.270	5	trailing zero is significant
0.00403	3	
270	2	
270.	3	decimal marks significance
270.0	4	
$4. \times 10^{-3}$	1	
4.00×10^{-3}	3	
$\overline{1}00$	2	with special notation to indicate 10's place, such as bold or overstrike

Identifying the number of significant digits in a number is relatively straightforward. Remember the following rules for your calculations:

1. **Multiplication / Division**: The least number of significant digits in any number in the equation, determines the number of significant digits in the answer.
2. **Addition / Subtraction**: The least number of significant digits in any number in the equation, determines the number of significant digits in the answer.

Make it a habit to check your significant digits before submitting your answer. Failure to report the final answer in the correct significant digits is considered a math error for grading purposes.

Useful Calculations

© Shutterstock.com

Making Dilutions

Standards are prepared by diluting from a known stock solution in two ways: 1) by carefully measuring required volumes from the stock solution, or 2) by sequential dilution.

To determine the appropriate amount of stock solution to add to make a new standard, use the following mass balance expression:

$$M_1V_1 = M_2V_2$$

Where:

- M_1 = concentration in solution #1
- V_1 = volume of solution #1
- M_2 = concentration in solution #2
- V_2 = volume of solution #2

A sequential (or serial) dilution can be used to accurately prepare very dilute standards (Figure 6).

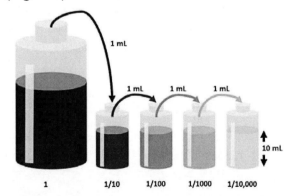

Figure 6. Depiction of a sequential (serial) dilution
Source: Daniel Meeroff

Both dilution methods are illustrated in the following example.

Given:
500 mg/L stock solution of NaCl

Find:
A) Describe how you would prepare 500 mL of a 5.0-mg/L standard by transferring NaCl directly from the stock solution

B) Describe how you would prepare 500 mL of a 5.0-mg/L standard, using 10-fold sequential dilutions

Solution:
A)
$$M_1V_1 = M_2V_2$$
$$(500\,mg/L)(V_1) = (5.0\,mg/L)(500\,mL)$$
$$V_1 = 5.0\,\text{mL}$$

So, measure out about 10 mL of the stock solution into a pre-labeled beaker. Carefully transfer 5.0 mL of stock solution from the beaker into a pre-labeled 500-mL volumetric flask (like the one shown in Figure 7). Bring the sample up to the 500-mL mark with distilled water and mix thoroughly.

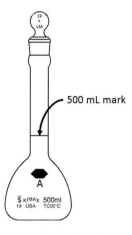

Figure 7. Location of the fill mark on a 500-mL Class A volumetric flask
Source: Daniel Meeroff

B) Place a little more than 50 mL of 500-mg/L stock solution into a pre-labeled beaker. Carefully transfer 50 mL of the stock solution from the beaker into a pre-labeled 500-mL volumetric flask. Bring the sample up to volume with distilled water and mix thoroughly. Then to a second pre-labeled 500-mL volumetric flask, transfer 50 mL of solution from the first volumetric flask to the second one. Bring the sample in the second flask up to volume with distilled water and mix thoroughly.

Note that the first volumetric flask has a concentration of 50 mg/L, which is a 10-fold dilution of the stock. The second volumetric flask has a concentration of 5.0 mg/L, which is a 10-fold dilution of the 50 mg/L solution in the first flask.

Reading the Meniscus

The water levels inside laboratory glassware are read at the bottom of the meniscus as shown in Figure 8.

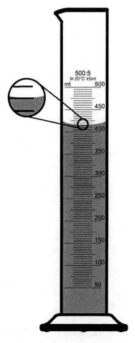

Figure 8. Location of meniscus (430 mL)
Source: Daniel Meeroff

To read the volume, bring the meniscus to eye level, either place your hand or a white index card behind it to increase the contrast, and then record the level of the meniscus.

Using a Pipetter

There are many different sizes of pipets in the lab. Make sure that are using the correct sized pipet and the correct pipet tips (size, sterile or non-sterile) for the task. To read the volume setting, look at the front face and you will see a window with 3 digits inside (Figure 9). You set the volume adjustment dial by turning the thumbwheel. Please take the time to learn how to read these properly to avoid damaging the dials or over-pipetting past its range.

P200   P1000

1		0
2		7
5		5

125 µL   0.75 mL

Figure 9. View of thumbwheel for P200 (left) and P1000 (right) pipetter
Source: Daniel Meeroff

A Simple Check for Proper Calibration

Check the calibration of your pipet using a known amount of deionized (or distilled) water to check if the output mass matches the setting.

For example, pipet 0.200 mL into a weigh boat on a top-loading balance having an accuracy level of at least a three decimal places (Figure 10). The mass should read $0.200 \pm 5\%$ grams.

Pipets having greater than 5% error should be recalibrated.

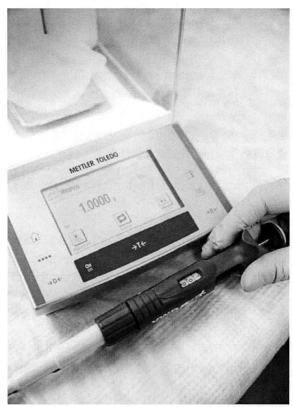

Figure 10. P200 calibration of 0.200 mL of distilled water
Courtesy of Florida Atlantic University

Pipetting Technique

1. GRIP: Hold the device by the body with your thumb finger on the plunger. Your index finger gives you much finer control over the plunger action. Do not hit the eject button accidentally.

© Shutterstock.com

2. LOADING: Load the appropriate sized tip

- Push the plunger down slowly to the point of first resistance; this sets the load volume.
- While holding the plunger at the load volume set point, put the tip into the solution so that it is immersed just enough to cover the end (3–4 mm).
- Slowly release the plunger to draw up the liquid, making sure to keep the tip immersed. Pause for a second before lifting the pipet out of the solution. Visually inspect the load to make sure it is correct and no air bubbles are observed.

3. DELIVERY: To deliver the volume, place the tip into the receiving vessel and press the plunger all the way to the second resistance point setting in one smooth motion. This expels all of the liquid. Then, without releasing the plunger, withdraw the tip.

Pre-Lab Responsibilities

Before attending the laboratory session, you must be prepared ahead of time for a safe and effective experience. A pre-lab quiz is required. If you score below the minimum, then you may not be allowed in the laboratory for that experiment.

First, **organize your time efficiently**. The best way is to **read the experiment** to become familiar with the procedure before the laboratory period. This way you will not have any surprises or wasted time during the session.

Second, **familiarize yourself with the details of the experiment**, including the standard operating procedures (SOPs), Material Safety Data Sheets (MSDSs), and safety information in order to be better prepared for any contingency.

Third, **plan ahead** so that you know approximately what will be done at each stage of the experiment. To derive benefits from the time and effort spent doing laboratory experiments, you should think about what the experiment is intending to illustrate. Become familiar with what will be happening during the session, why certain reagents are being used, and why things are done in a certain order. This will help maximize your time commitment.

© Shutterstock.com

Keep good lab notes and be observant. This will help in writing your lab report, and will also eliminate having to return at a later date to repeat the experiment. If you do not know what you will be doing until you set foot in the laboratory, you will be there a lot longer than you anticipated.

Laboratory Reports

Practitioners are looking to hire young engineers with excellent communication skills, and report writing is an important component of the engineering profession. Whenever you are writing a laboratory report, keep in mind that you must write accurately and with sufficient detail so that someone can replicate your experiment and results without your presence.

The format used for the laboratory report is applicable to many types of technical writing needed for most engineering careers. Copying sections of your lab report from the laboratory manual, other students (past or present), or any other source (e.g., the Internet) is considered plagiarism and will result in a zero grade.

Laboratory reports must be neat, clear, detailed, concise, well-organized, and professional in presentation. Hand-written lab reports are not acceptable. It is required that the final submittal be word processed and checked for spelling and grammar. It is likely that laboratory report will require re-writing and input from others in your group.

© Shutterstock.com

Consider yourself as an engineering consultant who is documenting analytical results for a very important wealthy client. You want to make a good impression because you want to continue doing business with this client.

It is strongly recommended that written laboratory reports be prepared immediately after completion of the experiment. Do not wait until the due date is upon you to begin writing the report.

Essential Elements

The elements included in the following checklist are required for all laboratory reports.

☐ **Title Page**
 1. Descriptive
 2. Include course title, experiment title, date of experiment, date of report, and names of team members
 3. Use key words (<15 words)

☐ **Table of Contents**

☐ **List of Figures/Tables**

☐ **Executive Summary** (also known as an "Abstract")
 1. Stands alone (< 200 words)
 2. Consists of 3 sections
    a. Concise description of objectives
    b. Methodologies
    c. Major findings

☐ **Introduction**
 1. Introduces the reader to the topic
 2. Explains the main concepts necessary to understand your experiment
 3. Discusses the basic theory and significance of the test procedure
 4. Refers to literature
 5. Defines the scope
 6. States objectives and hypothesis
 7. Explains how the experiment will address your objectives
 8. Ends with objectives

☐ **Methodology**
 1. List the steps in paragraph form as completely as possible
 2. Include detailed descriptions of equipment, materials, procedures, and quality assurance/quality control (QA/QC)
 3. Include a labeled photo of the experimental setup
 4. Mention analytical precision and standardization

☐ **Results and Discussion**
 1. Data collection and analysis using statistical tools
 2. Presentation of results in tabular or graphical format whenever possible
 3. Interpretation of results and brief discussion of each figure or table
 4. Discussion of correlations, trends, errors, or outliers observed

☐ **Conclusions**
 1. Restate the major findings
 2. Do not add any new information in the conclusions
 3. Make recommendations to limit error or improve the experiment

☐ **References**

☐ **Acknowledgments**

☐ **Appendices**

☐ **Blank Scoring Sheet**

Rubrics and Scoring Sheets are located in the back of this Laboratory Manual.

© Shutterstock.com

More about the Front Matter

Include experiment title, course title, instructor's name, date of experiment, date of report, and names of team members, and keywords.

```
Amount of Waste Collected in One Week for Four Households
Activity #7: Solid Waste Facility Field Trip and Waste Audit




                      ENV 3001C
           Environmental Science and Engineering
                  Daniel E. Meeroff, Ph. D.




             Date of Activity: July 28, 2017
            Date of Submittal: August 5, 2017




                   Team Gold Members:
                       Fred Salley
                        Jane Doe




                      Key Words:
recycling, waste audit, municipal solid waste, ferrous, non-ferrous, fiber, plastic
```

Source: Daniel Meeroff

Do you know how to make a List of Tables and a List of Figures?

First, make sure to add captions to all of your tables and figures using your word processor's automatic function:

- From the *References* menu, click *Insert Caption*.
- Pick the appropriate label, either *Figure* or *Table*.
- Then to make the list later, from the *References* menu, click *Insert Table of Figures*.

- To make the List of Tables, change the caption label to *Table*.

Follow the same procedure for your *Table of Contents*, except highlight your section headings, and change the heading style to *Heading 1*. After you have changed all of your headings, click where you want to add the Table of Contents. From the *References* menu, click *Table of Contents*.

You can update any table by clicking the table to highlight it and pressing F9.

More about the Abstract

The abstract is a three-paragraph summary of your lab report.

Paragraph 1. Contains introductory material, including a brief synopsis and objectives, and ends with the hypothesis.

Paragraph 2. Contains a brief description of each method used to carry out the objectives.

Paragraph 3. Contains a brief summary of major findings, including specific data results that prove or disprove the hypothesis. This paragraph may also contain a brief error discussion.

More about the Introduction

This section contains background material and basic theory to give the reader a starting point for understanding the experiment.

When writing the introduction section, here are a few tips that should help you:

1. Make an outline so you remember to cover all of the necessary basic concepts.

2. Write paragraphs that explain the basic concepts. Refer to the literature to help you with this. Be sure to describe the theory behind the procedures and your observations.
3. Now write a sentence that answers this question: "What is the objective (purpose) of this experiment?"
4. State your working hypothesis. In this statement include both your dependent and independent variables. Then give a rationale for your expectations and briefly mention the anticipated outcome.

More about the Hypothesis

A hypothesis can be tested in the lab, and it describes the anticipated outcome. This is the reason for conducting the experiment. A hypothesis must mention the condition that is being controlled in the experiment (*independent variable*) as well as the condition that is being measured in the experiment (*dependent variable*).

Here is a helpful example to illustrate the difference:

If identical potted plants are grown with either 1.0-L of distilled water or 1.0-L of concentrated sulfuric acid added for 10 days, then the leaf size of the distilled water plant will be larger, by at least 5%.

Independent variable (condition that you control) = <u>chemical added</u>.

Dependent variable (condition that you measure) = <u>leaf size</u>.

Now you try one.

Since male B. splendens have to protect fertilized eggs against predators, the aggressiveness of males, measured by the average number of displays per minute, will be higher in the presence of eggs than in the absence of eggs.

Note: this assumes that you explained the link between displays and aggressiveness in the introduction!

Now answer the following questions:

Independent variable =

_____

Dependent variable =

_____

More about the Methodology

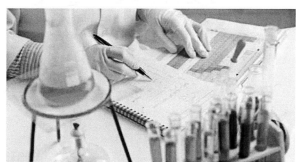
© Shutterstock.com

Make sure that enough information is provided so the experiment can be repeated by someone with basic laboratory knowledge.

Use the passive voice, past tense.

Example: The weight of the container was measured as 9.451 g.

The methodology section must contain a detailed description of the experimental setup. Take photographs or make a sketch of the setup (including details, dimensions, etc.). Imagine that you will need to recreate the setup for the final exam, so include enough detail. Address the following items in the methodology section:

- Provide the name, location, and general description of sampling sites, as well as the time sampled and what exactly was measured.
- Describe the steps in the order in which they were performed.
- Do not copy the procedure from the lab manual, since things change, and we may not always follow the lab manual.
- Provide details on the equipment (manufacturer and model number) used and its known accuracy.
- Describe the quality control procedures, and explain any differences between measured and expected values and accuracy checks.
- Describe how the experimental groups differ from the controls.
- Describe the measurements taken and what was recorded.
- Describe how you recorded and/or transformed the data.
- Describe any equations used and define all variables.

Example: The velocity (u) is estimated as $u = L/T$, where T is the time recorded for the sample to travel a distance, L.

Also describe your method of analysis, including what type of statistics were used, and how you made your comparisons.

Take a look at the following example to help you see how to do this better:

- **A t-test was calculated for the number of tail beatings. (Ambiguous!)**

- **A student's t-test was used to compare the number of tail beatings in male *B. splendens* with the number of tail beatings in females. (Much better!)**

What is quality assurance and quality control (QA/QC)?

QA/QC is best defined by citing an example. Say we are measuring temperature in the room with a digital thermometer. QA/QC could be to take several measurements with the same thermometer, or take similar measurements with another group's thermometer, or check the accuracy of the thermometer by dipping it in boiling water (which should read 100°C).

You just want to reassure yourself that the equipment is not faulty. This will increase your confidence level regarding your experimental results.

More about the Results and Discussion

This is the section in which you present your observations and results. It is difficult to do this if you failed to take good notes during the experiment. Be sure to record your observations during the experiment and complete your lab write-up as soon as possible after the experiment is concluded so that you do not forget any of the details.

Present the results prior to discussion or interpretation. Simply tell the reader what you observed by refer them in the text to the data tables.

Example: As shown in Table 2, a summary of the temperature measurements reveals that the temperature decreased over time.

Then follow up the table with statements that justify your analysis of the results and other evidence. Discuss what you have learned, and what trends the data may show. If there were no trends, but you thought there should have been, discuss that also.

© Shutterstock.com

In this section you should answer the following questions:

- Did the data correspond to what you expected? Why or why not?
- Were the results precise, accurate, too high, too low?
- Did the data correspond to what other people found (use citations)?
- If not, why do you think this happened? Present a new hypothesis.
- Discuss any problems that arose during the experiment. Unforeseen difficulties with the procedure may affect the data and should be described. List any weaknesses you identified in your experiment. You will need to tell the reader how they may have affected your results.
- Estimate the effects of these errors. Discuss the errors that may have shifted an otherwise reasonable answer in the direction of an erroneous one.
- Did the findings stimulate other questions for further research?

In many cases, you will have gathered data from several groups. When you share data or results, make sure you provide adequate annotation regarding who gathered which data, and discuss any deviations or peculiar observations. Confusion can be avoided if the results are presented neatly, with proper units, and with comments. Just in case others might be confused by your notes, provide a way for others to contact you for clarification (e.g., e-mail address). The instructor will not be responsible for getting lab data from other groups to you, so plan ahead.

More about the Figures & Tables

A picture is worth a thousand words, so use figures/tables/sketches to help present your findings and make your points. Discuss trends and comparisons in the text. Key graphs, figures, and tables should be embedded in the body of the report (not tacked on the end or buried in the appendix).

Use informative figure and table captions, and number them independently in the order that they are mentioned in the text. Raw data sheets belong in the appendix. For two-dimensional graphs, plot the independent variable on the x-axis (*abscissa*) and the dependent variable on the y-axis (*ordinate*). Figure captions are placed below the figure (as shown in Figure 11).

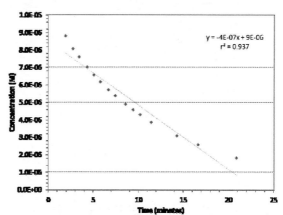

Figure 11. First order plot of concentration versus time for crystal violet sample #1
Source: Daniel Meeroff

Table captions are placed at the top of the table (as shown in Table 3). Remember that anything that is not a table is a figure.

Table 3. Summary of Water Quality Measurements Taken During November–December 2017 (Dania Beach, FL)

Site Location	Temp (°C)	Depth (ft)	Cond (mS/cm)	TDS (mg/L)
1	25.0–28.0	9	34.2–37.6	23.8
2	25.2–28.0	12	36.5–37.9	23.3
3	25.2–28.0	13	37.7–38.4	24.3
4	24.7–27.0	6	3.90–3.95	2.6

State your findings and rationale and then back them up with evidence (your evidence plus data from other research literature). Justify statements with results from statistical analyses and state if your hypothesis was supported or not by the observed data.

Watch the significant digits!

© Shutterstock.com

Be clear, be factual, and be specific. Following are some examples of phrases to avoid, and how to better discuss data.

Be quantitative about trends:

Weak Discussion: The data show a trend of decreasing dissolved oxygen with time.

Better Discussion: Dissolved oxygen decreased from 7.1 mg/L to 2.5 mg/L over a 5-day period.

Be specific and quantitative:

Weak Discussion: The data are scattered.

Better Discussion: The mean signal response was 3500 peak area units; however, the standard deviation was 4200 peak area units.

Take credit so long as the method was cited properly and accuracy was specified:

Weak Discussion: The laboratory clearly shows how alkalinity can be measured reliably.

Better Discussion: Because the relative error was 4.7%, we obtained an accurate measurement of alkalinity.

In terms of reporting, being correct is not sufficient. You must be correct and justify your findings.

As an example, consider the following three answers to the question:

"Did your data agree with the expected results?"

1. The laboratory results correlated with theory well.

2. As shown in Figure 1, the dissolved oxygen decreased over time. This was expected, as the microorganisms used the oxygen to degrade the organic matter. However, the oxygen increased on day 7, which was not expected.

3. As shown in Figure 1, the dissolved oxygen (DO) concentration started at 8.1 mg/L on day 0, which is 85% of the saturation value. The DO decreased to 0.8 mg/L on day 7. This was expected as the microbes used the oxygen to degrade the organic matter. However, the oxygen increased from day 6 to day 7, which was not expected. The method used to measure DO is only sensitive to 0.5 mg/L (APHA et al. 1998), and thus the increase may be an artifact of the method rather than a true increase.

Although all technically answer the question, it is clear that the depth of analysis is very different.

- The first response does not indicate any understanding of the results, as it provides only a yes/no answer.
- The second response provides some comparison to theory, but is still relatively qualitative.
- The third response provides more detailed comparison to theory, data analysis (for example, calculating the percent saturation of DO), and use of outside resources (for determining the method sensitivity).

More about the Discussion Questions

Discussion questions from the lab manual should be answered in a separate section following the results and discussion section. These questions are often designed to be answered using your experimental data from the lab exercise. If you do not have data from the experiment to justify your answer, you should use literature citations for support.

More about the Conclusions

© Shutterstock.com

This section should include a summary of the major results, a brief sentence about data quality, and a list of possible errors that could have occurred during the experiment. In addition, recommendations should be put forward to address ways to improve the experiment to gain more useful information or more accurate data. Make sure that your conclusions follow logically from your results and discussion.

Begin the conclusion by restating your hypothesis and supporting or refuting it with your experimental results. Identify any errors and their impact on the results. Then provide suggestions on how to avoid these errors.

Discuss the broader impacts of the results. In other words, how your findings will be used in your professional career. Finally, suggest ways to improve the laboratory experience or mention other experiments that you are inspired to perform after conducting this one.

More about the References

Typically, you should cite a minimum of 5 references. **No more than 2 Internet sources are allowed.** Following are a couple of ways to cite references correctly in the text.

Example 1.
Methoprene has been shown to significantly affect the development of *Drosophila melanogaster* (Jones 2003).

Example 2.
Jones (2003) has found that methoprene significantly impacts the development of fruit flies.

Cite all references used. Avoid copying and pasting from the Internet or someone else's work. This is called **plagiarism** and is grounds for dismissal from the university. When using reference material in the text, always paraphrase (try to avoid direct quotes). However, if you do use text directly, make sure the statements are properly quoted and cited. In this course, we will use the citation format of the *Journal of Environmental Engineering* published by the American Society of Civil Engineers (ASCE; http://pubs.asce.org).

Examples of Reference Format

Journal References
Include year, volume, issue, and page numbers.

Stahl, D. C., Wolfe, R. W., and Begel, M. (2004). "Improved analysis of timber rivet connections." *Journal of Structural Engineering*, 130(8), 1272–1279.

Conference Proceedings and Symposiums
Include the conference sponsor or publisher of the proceedings, AND that entity's location—city and state or city and country.

Garrett, D. L. (2003). "Coupled analysis of floating production systems." *Proceedings of the International Symposium on Deep Mooring Systems*, ASCE, Reston, Va., 152–167.

Books
Include author, book title, publisher, the publisher's location, and chapter title and inclusive page numbers (if applicable).

Zadeh, L. A. (1981). "Possibility theory and soft data analysis." *Mathematical frontiers of the social and policy sciences*, L. Cobb and R. M. Thrall, eds., Westview, Boulder, Colo., 69–129.

Reports
Use the same format as that shown for books. For reports authored by institutions, spell out the institution acronym on first use, and follow with the acronym in parentheses, if applicable. For reports authored by persons, include the full institution name—no acronym—and its location.

Unpublished Material
Unpublished material is not included in the references but may be cited in the text as follows: (John Smith, personal communication, May 16, 1983); or (J. Smith, unpublished internal report, February 2003).

Web Pages
Include author, copyright date, title of "page," Web address, and the date on which the material was downloaded.

Burka, L.P. (1993). "A hypertext history of multi-user dimensions." *MUD history*, <http://www.ccs.neu.edu> (Dec. 5, 1994).

Theses and Dissertations
Include authors, copyright date, title, and the name and location of the institution where the research was conducted. Note that some institutions use specific terminology—for example, "doctoral dissertation" rather than "PhD thesis."

Sotiropulos, S. N. (1991). "Static response of bridge superstructures made of fiber reinforced plastic." MS thesis, West Virginia Univ., Morgantown, WV.

Provide a screenshot of the first page of any article or book that you cited and include that in the appendix so the quality of your references can be evaluated.

More about the Appendices

Attach a copy of your raw data (correctly arranged and labeled), the outputs for any statistical tests you ran, extra photographs not used in the body of the lab report, supporting documents (e.g., MSDS), and the first page of each scientific paper or textbook you cited.

General Format

Make sure that your report is typed with 1-inch margins all around, block justified. Use 1.5-line spacing and use Times or Arial 11-pt. font. Here are some additional helpful hints to make your laboratory report look as professional as possible:

- Numbers must always include units (Ex. 3.0 mg/L, not 3).
- If <1, always have a 0 in front (Ex. 0.1, not .1).
- If ≤5 and no units, then spell the number out, and if >5, then use the number. For example, a group of four vials; a total of 12 experiments.

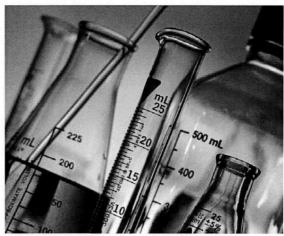
© Shutterstock.com

Do not use informal (non-standard) abbreviations in your paper (such as "temp." for temperature). Write using complete sentences, and avoid contractions. Organize your report into clear, cohesive sections rather than one long block of text.

Writing well takes time, practice, and patience. You cannot write a masterpiece the night before the assignment is due. Furthermore, because quality is part of your grade, you should endeavor to get feedback and comments prior to turning in your work. **Therefore, do the lab write-up SOON after doing the lab!** If you wait until the last minute, you may not be proud of the quality of the work that you turn in.

Trip Reports

The goal of writing a field trip report is to enhance your understanding of real-world integration of engineered systems with the natural environment and of course to continue to practice your technical writing skills.

© Shutterstock.com

You may find it easier if you try to tell the story of what happened on your trip. The purpose of the trip report is to document your impressions gained as a result of your experiences, together with the basic facts and details of the destination.

Essential Elements

Here is a list of some of the important areas you will need to cover:

- ☐ Where exactly did you go?
- ☐ How did you get there?
- ☐ When did you go?
- ☐ What was the purpose of the visit?
- ☐ Detailed descriptions of what you observed
- ☐ Were there any difficulties or local hazards to note?
- ☐ What was the weather like? Is that typical?
- ☐ How does this location compare with other similar locations to which you have been?
- ☐ Questions that were or were not answered during the trip
- ☐ Recommendations on how to improve the quality of the experience
- ☐ **Acknowledgements**: At the end of the report, at the end of the report provide the full name and job title of tour operators or others who assisted during the field trip.

Photography

Digital photographs can help capture the experience and provide illustrations for deeper understanding. However, you should be selective; dozens of pictures of the same thing will be tedious and detract from the quality of the document. Ideally, photos included in the report should illustrate the location, important facilities, engineered systems, wildlife (flora and fauna), and action shots of tour guides with the group.

Photographs do not have to be your own (particularly if you do not own a camera), but always give full credit to the photographer in the report. You should incorporate the photographs in the text of the report as figures.

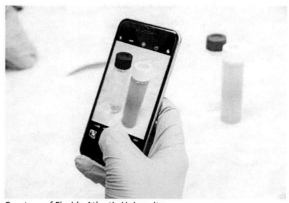

Courtesy of Florida Atlantic University

Now you are ready to enjoy your first laboratory experience!

For your convenience, each activity provides an introductory summary of the concepts to be studied. This helps to fill in the important background material and basic concepts for less experienced students and provides a useful review for the more experienced ones.

Open your mind and allow yourself to discover the wonders of experimental observation. This will help you to better understand the world around you and give you the tools to explore the unknown.

So grab a pen, paper and PPE, and let's go!

© Shutterstock.com

Activity #1. pH

Purpose:

- To determine the pH of natural water samples and observe differences in pH measurement methods.

Background

The pH is the negative log of the hydrogen ion concentration:

$$pH = -\log[H^+]$$

An *acid* donates H^+ (or accepts OH^-), while a *base* accepts H^+ (or donates OH^-).

- Acidic waters have a pH < 7
- Basic waters have a pH > 7
- Neutral waters have a pH = 7

The pH parameter is typically measured with a pH electrode and a pH meter. These instruments must first be calibrated using freshly prepared standard buffer solutions.
The typical pH electrode is a combination (glass pH electrode) containing either a calomel ($Hg/HgCl_2$) or $Ag/AgCl$ reference electrode. The meter measures the response from the electrode as an electrical potential

Aquatic organisms (fish, plankton, algae) depend on circumneutral pH (between 6 and 9). Also, toxic metal concentrations can greatly increase as pH lowers through the dissolution of metal hydroxides.

Pre-Lab Questions

1. Why is pH important in civil engineering?

2. Name one kind of sample would you **not** be able to use a pH meter for, and explain why.

3. Provide a table that shows your **expected pH values** for all of your environmental samples. Provide a reference for your expected values.

4. If atmospheric CO_2 is predicted to increase from 350 ppm to 720 ppm in our lifetimes, what will be the new pH of natural rainwater? Show your work, list your assumptions, and check your assumptions for validity.

5. According to the MSDS, what kind of PPE should you use in this lab?

Procedure

1. **Obtain two samples.** Each group should provide **two** environmental samples. You need to bring **~500 mL** of each sample.

- One sample should be **rain water**. If you cannot obtain rain water, try a surface water (e.g. lake water, canal water, etc.).

- Make sure that your sample bottle is collected with no head space (completely full) and remains refrigerated until the day of the experiment.

- **Note:** Do **NOT** bring swimming pool water, bottled water, soda, seawater, artificially colored beverages, wine, beer, Windex, coffee, tea, milk.

- One QA/QC sample will be provided by the TA.

- Bring all samples to room temperature before analysis.

2. **Calibrate the pH meter, if necessary.** If the meter is already calibrated skip to step #3. Otherwise, calibrate the instrument per the manufacturer's instructions.

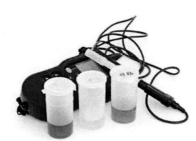

© Shutterstock.com

- Between samples or calibration standards, you must use deionized (or distilled) water to rinse the electrode.

- Always catch the rinsate in a waste beaker.

- Always blot the probe dry with a clean Kimwipe before placing the probe in the next standard or sample.

- Make sure that the probe is always submerged in liquid between readings. Do not let it get dry.

3. **pH Test Strips.** Take one pH test strip being careful not to touch the color squares, and dip the strip into the sample until the color no longer changes (~5 seconds). Pull out the strip, and immediately compare the color to the color standards on the package label. Repeat twice for all samples, and record the values in your data table in the column labeled, "pH test strip."

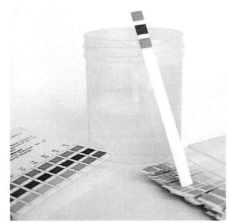

© Shutterstock.com

4. **Measure the pH and temperature of your sample(s).** Stir each solution with a stir bar to achieve a stable reading. Follow the same procedure for rinsing and drying the probe between samples as described in step 2, and repeat for all three samples. **Measure your pH values at least 2 times.**

5. **Tabulate your results.** Provide a table just like the one that follows:

Sample	Trial #	pH test strip	pH meter	°C
QA/QC	1			
QA/QC	2			
Rain	1			
Rain	2			
Canal	1			
Canal	2			

Discussion Questions

1. Provide a table that shows your three samples with expected pH values, pH paper values, measured pH values from the probe, temperatures, and percent errors between the pH strip and the pH probe readings. In this experiment, you took multiple probe readings and multiple test strip readings, so report your range (minimum and maximum), average, and standard deviation values for each sample. Discuss why there are differences between the pH meter and the pH test strip.

2. Find out the actual value of the QA/QC accuracy check sample from the TA, and calculate your percent error for the pH test strip and for the pH meter.

3. Which method (pH strip or pH meter) is more accurate, which is more precise?

Proper Safety and Disposal

Do NOT pour your waste into the sink!

Only samples that have a pH between 6 and 9 can be dumped into the sink. All others must be neutralized by the TA.

Empty your glassware into the proper waste container, and wash thoroughly with soap and water. Then rinse with deionized water and place on the drying rack.

If you fail to follow these instructions, then you may have to stay after time is up.

Activity #2. Alkalinity

Purpose:

- To determine the total alkalinity, OH⁻ alkalinity, CO_3^{2-} alkalinity, and HCO_3^- alkalinity of natural water samples.

Background

Alkalinity is defined as the amount of acid required per liter of solution to lower the pH to less than 4.3. Essentially, it is a measure of the capacity for a solution to neutralize strong acid.

Natural waters are buffered by the *carbonate system*:

$$CO_{2(aq)} + H_2O_{(l)} \longleftrightarrow H_2CO_{3(aq)}$$

$$K = 10^{-1.47} \, M/atm$$

$$H_2CO_{3(aq)} \xleftrightarrow{K_{a1}} H^+_{(aq)} + HCO_{3\,(aq)}^-$$

$$K_{a1} = 10^{-6.37} \, M$$

$$HCO_{3\,(aq)}^- \xleftrightarrow{K_{a2}} H^+_{(aq)} + CO_{3\,(aq)}^{2-}$$

$$K_{a2} = 10^{-10.33} \, M$$

The alkalinity equation is derived from a simple charge balance on the four major species in a natural water system. The total alkalinity in units of eq/L is expressed as:

Alkalinity (eq/L) = [HCO₃⁻] + 2 [CO₃²⁻] + [OH⁻] − [H⁺]

Note that the symbol "[ ]" refers to the molar concentration, but you already knew that! In engineering applications, alkalinity is expressed in terms of mg/L as $CaCO_3$, which is computed as follows.

Alkalinity (mg/L as CaCO₃)
= Alkalinity (eq/L) × 50 g CaCO₃/eq × 1000 mg/g

The dominant species in the carbonate system is dictated by the pC-pH diagram (Figure 12), which shows the relative abundance as a function of pH.

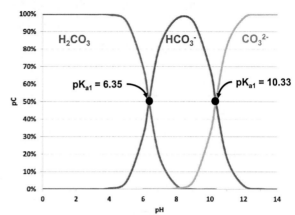

Figure 12. pC-pH diagram for the carbonate system
Source: Daniel Meeroff

For example, if the pH of a natural water sample is 6.9, Figure 12 tells us that the major species in this range is bicarbonate (HCO_3^-), thus the alkalinity under these circumstances is approximately equal to the bicarbonate concentration.

We can determine the alkalinity of a solution by titrating with a strong acid (H_2SO_4) in the presence of a colored indicator. You have probably titrated before using a buret, but you can also use a digital titrator (Figure 13).

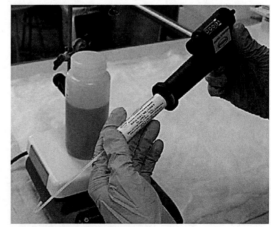

Figure 13. Digital titrator
Source: Daniel Meeroff

The indicators for the end points will be phenolphthalein and bromocresol green-methyl red. The phenolphthalein indicator changes from pink to colorless at a pH around 8.3. This is an indication of the OH^- alkalinity plus one half of the CO_3^{2-} alkalinity (see Table 4).

The bromocresol green-methyl red indicator changes color from blue to salmon at a pH around 4.3. This is an indication of one half of the CO_3^{2-} alkalinity plus the HCO_3^- alkalinity (see Table 4).

The result of the titration should look like the data plotted in Figure 14. The x-axis corresponds to the volume, in mL, of the titrant (in this case H_2SO_4) added.

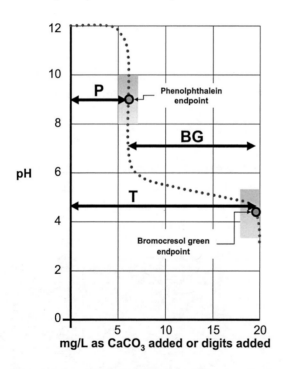

Figure 14. Theoretical titration curve showing the location of the phenolphthalein and bromocresol green-methyl red endpoints
Source: Daniel Meeroff

The total alkalinity is the sum of the titratable bases: caustic alkalinity, carbonate alkalinity, and bicarbonate alkalinity.

Caustic Alkalinity (OH^- Alkalinity) is the moles of strong acid required to lower the pH of the sample to consume all the hydroxide.

$$OH^-\ Alk\ (eq/L) = [OH^-] - [H^+] - [HCO_3^-] - 2[H_2CO_3]$$

Carbonate Alkalinity (CO_3^{2-} Alkalinity) is the moles of strong acid required to lower the pH of the sample to consume all the carbonate (pH~8.3).

$$CO_3^{2-}\ Alk\ (eq/L) = [CO_3^{2-}] + [OH^-] - [H^+] - [H_2CO_3]$$

The amount (eq/L) of H_2SO_4 needed to reach the phenolphthalein endpoint (pH~8.3) is called the **phenolphthalein alkalinity (P)**. The additional amount of H_2SO_4 used to reach the bromocresol green-methyl red endpoint (pH~4.3) is known as the **bromocresol green alkalinity (BG)**. The two quantities added together represent the **total alkalinity (T)**. All alkalinity computations are summarized in Table 4.

Table 4. Alkalinity computation table

Result	OH^- Alkalinity	CO_3^{2-} Alkalinity	HCO_3^- Alkalinity
1. Phenolphthalein Alk = 0	= 0	= 0	= T
2. Phenolphthalein Alk = Total Alk	= T	= 0	= 0
3. Phenolphthalein Alk < 0.5 × Total Alk	= 0	= 2P	= T – 2P
4. Phenolphthalein Alk = 0.5 × Total Alk	= 0	= T	= 0
5. Phenolphthalein Alk > 0.5 × Total Alk	= 2P – T	=2(T – P)	= 0

To titrate efficiently, you must use the appropriate sample volume, the appropriate titration cartridge, and the appropriate digit multiplier, as shown in Table 5.

Table 5. Appropriate volumes, cartridges, and digit multiplier for digital alkalinity titrations

Range (mg/L as $CaCO_3$)	Sample Volume* (mL)	Titration Cartridge (H_2SO_4)	Digit Multiplier
10 – 40	100	0.1600 N	0.1
40 – 160	25	0.1600 N	0.4
100 – 400	100	1.600 N	1.0
200 – 1000	50	1.600 N	2.0
1000 – 4000	10	1.600 N	10.0

*Remember the final volume will always be 100 mL

Example 1.1. A sample has an expected alkalinity of 2500 mg/L as $CaCO_3$, then according to Table 5, what sample volume, which cartridge should you use, and what is the digit multiplier you should apply?

Given:
Expected alkalinity = 2500 mg/L as $CaCO_3$

Find:
Which sample volume, cartridge, and digit multiplier to use?

Solution:
According to Table 5, a sample with 2500 mg/L as $CaCO_3$ corresponds to the last row. Therefore, you should use 10 mL of sample diluted to 100 mL with 90 mL of deionized water, and you should use the 1.600 N cartridge. The final number of digits will then be multiplied by 10.0 to account for the dilution. If you went with the smaller cartridge (0.1600 N), then you would be titrating for a very long time before reaching the endpoint!

Let's do an example to make sure you know how to compute the different alkalinities using Table 4 and Table 5.

Example 1.2. A 100 mL sample takes 175 digits of 0.1600N H_2SO_4 to reach the phenolphthalein endpoint and another 85. digits to reach the bromocresol green-methyl red endpoint. What is the concentration of hydroxide, carbonate and bicarbonate alkalinities?

Given:
175 digits to reach P, 85 digits to reach BG using the 0.1600 N cartridge with 100 mL sample

Find:
Concentration of hydroxide, carbonate, and bicarbonate alkalinities = ?

Solution:
Using Table 5, we find that the multiplier is 0.1 from the first row, so we get:

$$P = 175 \times 0.1 \quad = 17.5 \text{ mg/L as } CaCO_3$$
$$BG = 85 \times 0.1 \quad = 8.5 \text{ mg/L as } CaCO_3$$
$$T = P + BG \quad = 26.0 \text{ mg/L as } CaCO_3$$

Step 1. P ≠ 0 (it is 17.5 mg/L). Go to row 2 (Table 4).

Step 2. P ≠ T (17.5 mg/L vs. 26.0 mg/L). Go to row 3 (Table 4).

Step 3. The total alkalinity divided by 2 = 13.0 mg/L. Since 17.5 mg/L > 13.0 mg/L, we will use the last row in Table 4 (row 5).

Step 4. Compute from row 5 in this case:

- OH^- alkalinity = 2P – T
  = 35.0 – 26.0
  = 9.0 mg/L OH^- alkalinity

- CO_3^{2-} alkalinity = 2(T – P)
  = 2 (26.0 – 17.5)
  = 17.0 mg/L CO_3^{2-} alkalinity

- HCO_3^- alkalinity = 0
  = 0.00 mg/L HCO_3^- alkalinity

Step 5. Do a quick check:
The sum of the three individual alkalinities must equal the total alkalinity.

	9.0 mg/L hydroxide alkalinity
+	17.0 mg/L carbonate alkalinity
+	0.0 mg/L bicarbonate alkalinity
=	**26.0 mg/L as CaCO₃ = T** ☑

Remember: *You must do this same calculation using your data collected in this experiment.*

Pre-Lab Questions

1. Why is alkalinity important in civil engineering?

2. Provide a table that shows your **expected alkalinity values** for all of your environmental samples. Provide a reference for your expected values.

3. If your sample has a pH of 8.90, what color will the solution be after you add the phenolphthalein?

4. You dilute your sample by taking 20.0 mL of canal water and mix with 80.0 mL of deionized water to make a total volume of 100.0 mL, using the 1.600 N cartridge, you reach the endpoint after 99.0 digits, what is the total alkalinity of your sample in mg/L as CaCO₃?

5. A sample is diluted by taking 10 mL and mixing with 90 mL of deionized water for a total of 100 mL. This solution takes 75 digits of 1.600N H_2SO_4 to reach the phenolphthalein endpoint and another 55 digits to reach the bromocresol green-methyl red endpoint. What is the total alkalinity and what is the concentration of hydroxide, carbonate, and bicarbonate alkalinities in the original sample?

6. According to the MSDS, what kind of PPE should you use in this lab?

Procedure

1. Use the same two environmental samples from Activity #1. One additional QA/QC sample for alkalinity will be provided by the TA.

2. **Determine the expected alkalinity.** For each of your samples, take one alkalinity test strip being careful not to touch the color squares, and dip the strip into the sample for 1 second. Pull out the strip, and compare the color to the color standards on the package label, within 30 seconds (see Figure 15). This is the ***expected alkalinity***. Make sure to record this value in your data table.

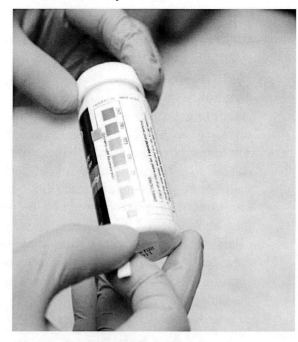

Figure 15. Alkalinity test strip
Courtesy of Florida Atlantic University

3. **Select the appropriate values from Table 5.** The sample volume, titration cartridge, and digit multiplier are determined from the corresponding *expected alkalinity* just like in Example 1.1.

4. **Prepare the digital titrator.** Slide the appropriate cartridge into the titrator receptacle and lock into position with a slight turn. Remove the polyethylene cap and insert a clean delivery J-tube into the end of the cartridge until it is tight. To start titrant flowing, hold the tip of the cartridge up. Advance the plunger release button (push the button in and toward the cartridge) to engage the piston with the cartridge. Turn the delivery knob until air is expelled and several drops of solution flow from the tip. Then use the counter reset knob to return the digital counter back to zero, and wipe the tip before use.

4. **Prepare the sample.** Start with the alkalinity QA/QC sample. Use a graduated cylinder or pipet to measure the appropriate sample volume indicated from Table 5 just like in Example 1.1

5. **Transfer the sample into a clean beaker.** **Record the initial pH and temperature of the sample**. Then reset the digital titrator counter to zero.

6. **Add the phenolphthalein indicator.** Add the contents of one phenolphthalein powder pillow to the sample. Stir to mix with the magnetic stir bar. *If the solution is colorless after addition of phenolphthalein indicator, then skip to step 8 because the phenolphthalein alkalinity is equal to zero.*

7. **Evaluate phenolphthalein alkalinity. If a pink color appears, add titrant (H$_2$SO$_4$) slowly, to the colorless endpoint.** Make sure to place the delivery tube tip into the solution and swirl the flask or stir while titrating. Record the pH readings after every few digits and every digit near the endpoint (pH~8.3). **Record the number of digits on the titrator (or better yet, record both initial and final readings) in a table**. Mark the endpoint clearly on your table.

Sample Vol. (mL)	Cartridge (M)	pH	Temp. (°C)	Digits	Multiplier	Color

8. **Add the bromocresol green-methyl red indicator.** Add the contents of one bromocresol green-methyl red powder pillow. Stir to mix with the magnetic stir bar. The color should be blue-green.

9. **Evaluate bromocresol-green alkalinity.** Reset the titrator, and titrate from blue-green to the final endpoint indicated in Table 6, checking the pH with a pH meter. Do not over-titrate!

Table 6. Typical alkalinity titration endpoints

Sample composition	Endpoint	Color
Alkalinity ~30 mg/L	pH 4.9	Light violet or gray
Alkalinity ~150 mg/L	pH 4.6	Light pink
Alkalinity ~500 mg/L	pH 4.3	Light pink
Silicates/Phosphates present	pH 4.5	Light pink
Complex matrix	pH 4.5	Light pink

10. **Record your pH, temperature, and color readings in a table.** Regardless of the sample, if pH < 4.3, the endpoint has been reached (**record your readings**).

11. **Plot one titration curve.** For your QA/QC sample only, plot a titration curve as shown in Figure 14.

12. **Repeat the titration for your other two samples.** For the other two samples, you can go directly to the endpoint, without

taking pH measurements and plotting a titration curve.

13. **Perform all calculations to determine T, P, and BG from all your samples.** The calculations should follow Example 1.2. Tabulate all your pH values, digits, and alkalinities for each sample. *Be sure to provide a breakdown of total, carbonate, and caustic alkalinities for each sample.* Note: if you forgot to reset the titrator after P alkalinity, then you will be reading the Total alkalinity (T) instead of BG alkalinity. Remember these are related by the following equation:

$$T = P + BG$$

Discussion Questions

1. Provide a table that shows all samples with expected alkalinity values (from your pre-lab), alkalinity paper values, measured digits, digit multiplier, measured digital titrator alkalinity, and percent errors between the strip and titrator readings. Discuss why there are differences between the digital titrator and the alkalinity test strip.

2. What do you do if you want to conduct an alkalinity titration and your sample has its own pink color before the addition of the phenolphthalein indicator?

3. Can an acidic solution have a measurable alkalinity? Justify your response using an example from your data.

Proper Safety and Disposal

Do NOT pour your waste into the sink!

Only samples that have a pH between 6 and 9 can be dumped into the sink. All others must be neutralized by the TA. That includes all of the alkalinity samples.

Empty your glassware into the proper waste container, and wash thoroughly with soap and water. Then rinse with deionized water and place on the drying rack.

If you fail to follow these instructions, then you may have to stay after time is up.

Activity #3. Petri Dish

Purpose:

- To determine the number and types of microorganisms in environmental samples using the heterotrophic plate count (HPC) method.

- To determine the accuracy of HPC using the spread plate technique for different dilutions.

- To determine the growth rate of typical bacterial colonies on HPC plates.

Background

The **heterotrophic plate count (HPC)**, also known as the standard plate count, is a common procedure for estimating the number of live microorganisms in water, in air, or on surfaces. It measures the number of colony forming units that arise after a short incubation period on growth media in a Petri dish. Colonies of microorganisms arise from pairs, chains, clusters, or single cells. All of these are included in the term *"colony-forming units"* (CFU).

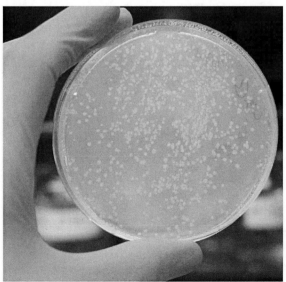

© Shutterstock.com

In practice, several different methods for performing the heterotrophic plate count are used.

a) Spread plate
b) Pour plate
c) Membrane filtration

Recently, new methods developed by IDEXX utilize a multiple enzyme technology. These tests enumerate the microorganisms with an estimate called the **most probable number** (MPN), which is not necessarily equivalent to CFU.

For this experiment, we will employ the *spread plate method*, as shown in Figure 16 as well as the IDEXX Quanti-Tray*/2000 procedure.

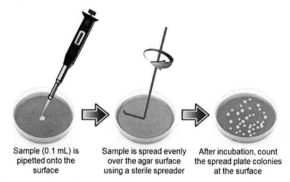

Sample (0.1 mL) is pipetted onto the surface

Sample is spread evenly over the agar surface using a sterile spreader

After incubation, count the spread plate colonies at the surface

Figure 16. Diagram summary of the spread plate technique
Source: Daniel Meeroff

For the spread plate method, we will inoculate a Petri dish with microorganisms from an environmental sample using a measured quantity (typically 0.1 mL or less). We will distribute the liquid evenly across the surface of the media using a sterile L-spreader, and then incubate the plate for 48 hours at the appropriate temperature. At this point, we will count the number and diversity of microbial colonies visible on the plate. Using this number, we can calculate the number of microorganisms present in the original sample.

HPC values are reported as CFU per mL or per gram. Plates with counts between 20 and 300 colonies (the ideal range) are the most accurate. Those plates with counts that exceed 300, should be labeled as "**TNTC**," which stands for "**too numerous to count**." Those plates with less than 20 colonies should be reported as "**< 20**."

Results are reported by specifying the incubation time, temperature, method, and type of media, for example:

Heterotrophic Plate Count at 35°C/48 hours = 120 CFU/mL. This test was performed by the spread plate method on R2A Agar.

When we do not know how many microbes are in a sample before we start, we make *serial dilutions*, and use only those plates with colony counts between 20 and 300. Serial dilutions are described in Figure 17.

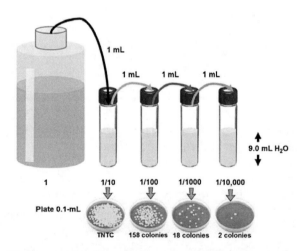

Figure 17. Example of serial dilution procedure
Source: Daniel Meeroff

In the example in Figure 17, all four vials initially contain 9.0 mL of sterile dilution water, and 1.0 mL of the previous dilution is added to the vial to make the final volume 10 mL. In the example, the only Petri dish that is countable is the 1/100 dilution. The 1/1000

and 1/10,000 plates are less than 20, and the 1/10 is too numerous to count.

So our HPC value in the original sample will be calculated by taking the colony count divided by the dilution factor and then dividing by the amount of liquid inoculated on the dish, as follows:

$$HPC = 158 \div \frac{1}{100} \times \frac{1}{0.1\,mL} = 1.58 \times 10^5\ CFU/mL$$

After the plates are incubated, sometimes we will observe a monoculture or multiple colony types (Figure 18).

0 CFU/mL	42 CFU/mL	42 CFU/mL
0 colony types	4 colony types	1 colony types

Figure 18. Example of monoculture (right) and multiple colony types (middle)
Source: Daniel Meeroff

If too many colonies (>300) are present, the whole plate does not have to be counted; rather, the count can be estimated by the quadrant technique, as follows:

1. Divide the dish into quadrants,
2. Count the number of CFU in one of the countable quadrants,
3. Determine the total estimated plate count by multiplying the partial count by four,
4. The estimated plate count (CFU) is then divided by the sample volume (mL) to give the estimated HPC (CFU/mL).

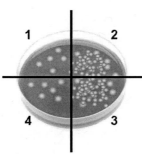

Source: Daniel Meeroff

In the example shown here, quadrants 2 and 3 are not countable, so counting quadrant 4 yields 6 × 4 = 24 CFU on this plate, and counting quadrant 1 yields 5 × 4 = 20 CFU on the plate. In this example the average estimated plate count would be 22 CFU.

The quadrant technique is also useful if you have smears or overgrowth ("lawn"). If this happens, simply draw your quadrants so that one of them falls on a countable portion of the plate, and report the value as an estimated value, just like if it had too many colonies.

To compute the HPC and report the value as CFU/mL, the total number of colonies or average number per plate (if counting multiple plates of the same dilution) is adjusted by the dilution factor.

Let's do an example on how to report the HPC value correctly.

Example 3.1. A sample has 39 total colonies (red circles and white creamy colonies) on R2A Agar after 48 hours incubation at 35°C, which was inoculated with 0.1 mL of the 1/100 dilution.
What is the reported HPC value?

First, compute the numerical value by dividing the colony count by the dilution factor (1/100), then adjust for the inoculation volume of 0.1 mL:

$$HPC = 39 \div \frac{1}{100} \times \frac{1}{0.1} = 3.9 \times 10^5 \ CFU / mL$$

- *Heterotrophic Plate Count at 35°C/48 hours = 3.9×10^5 CFU/mL, this test was performed by the spread plate method on R2A Agar.*

You must report the value of all of your plates using this same method.

Sometimes mold and fungi are present when the HPC is counted (Figure 19).

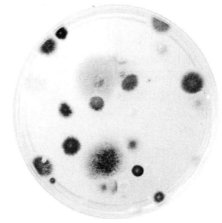

© Shutterstock.com

Figure 19. Mold and fungi growing on the Petri dish

If there are a few isolated colonies of mold on the agar plate, NOT interfering with the reading of the colonies, the number of mold are not reported, but their presence is noted, as in the example that follows:

Ex. If 45 colonies were counted (0.1 mL of sample tested), the HPC result would be:

- *Heterotrophic Plate Count at 35°C/48 hours = 4.5×10^2 CFU/mL, this test was performed by the spread plate method on R2A Agar. **NOTE: Fungi also present.***

In cases where mold growth interferes with counting bacterial colonies, the count is reported with the comment as follows:

Ex. If the count is 45 colonies (0.1 mL of sample tested) and there is evidence that there are other colonies under heavy mold growth, the HPC is reported as:

- *Heterotrophic Plate Count at 35°C/48 hours: 4.5×10^2 CFU/mL, this test was performed by the spread plate method on R2A Agar. **NOTE: Fungal growth in the sample may result in an incorrect estimate of the actual heterotrophic plate count.***

If **only** mold is growing on the plate (0.1 mL of sample tested), that is, there is no visible growth of bacteria, the HPC is reported as:

- *Heterotrophic Plate Count at 35°C/48 hours = less than 20 CFU/mL, this test was performed by the spread plate method on Plate Count Agar. **Note: Fungi present.***

Pre-Lab Questions

1. Provide two examples of engineering applications which involve enumeration of HPC and disinfection?

2. What are heterotrophic bacteria? Give three specific microorganism names included in this group.

3. What is the total microbial count that you expect in each of your environmental samples in CFU/mL?

4. If you count 25 colonies after 48 hours incubation on a plate inoculated with 0.1 mL of a 1/1000 dilution, what is the HPC in CFU/mL of the original sample?

5. How many colonies would you expect on the plate, if the dilution above was 1/10,000?

6. List the ingredients in R2A Agar.

Procedure

1. **Obtain one sample.** Each group should provide **one** environmental sample. You will need to use less than **10 mL** of sample (i.e. river water, rain water, sea water, well water, or canal water will work as long as it is not chlorinated).

- **Do not bring in water from the tap, which contains chlorine bleach.**

- *Do NOT clean your sampling container with bleach.*

- Collect your sample less than 4 hours before the lab time and keep refrigerated for best results. Remember to bring to room temperature before making your serial dilutions.

2. **Prepare Spread Plates.** Each group will be furnished with a number of Petri dishes for spread plate testing (see Figure 20). *If your plate has condensation on it, do not turn it over until you remove the top lid and wipe off the excess water with a Kimwipe.*

Figure 20. Spread plate testing station
Source: Daniel Meeroff

3. **Prepare Serial Dilutions.**

- **First serial dilution (1/100).** Using a P200 pipet with sterile tip, measure exactly 0.1 mL of sample into a vial containing exactly 9.9 mL of sterile dilution water.

- $Dilution = \dfrac{0.1\ mL}{(9.9\ mL + 0.1\ mL)} = \dfrac{1}{100}$

- Close the cap (label this vial as the 1/100 dilution) and mix well. Discard this pipet tip. Remember the pipet tips are sterile, the vial is sterile, but the cap is not, so do not invert when mixing.

- **Second serial dilution (1/1000):** Using a P1000 pipet, measure exactly 1.0 mL of the 1/100 dilution into a new vial containing exactly 9.0 mL of sterile dilution water.

- Close the cap (label this vial as 1/1000 dilution) and mix well. Discard this pipet tip.

- **Third serial dilution (1/10,000):** using a P1000 pipet, measure exactly 1.0 mL of the 1/1000 dilution into a third vial containing exactly 9.0 mL of sterile dilution water.

- Close the cap (label this vial as 1/10,000 dilution) and mix well. Discard this pipet tip.

4. **Label all plates.** Write small (along the rim) so as not to obscure the viewing area (Figure 21).

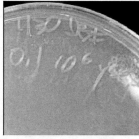

Figure 21. Proper technique for labeling Petri dishes
Source: Daniel Meeroff

Only label the **bottom lid** of all Petri dishes with your group name, date, the dilution, and the sample identification.

5. **Inoculate Plates.** Lay all of your Petri dishes on a level surface with the lid side up. **DO NOT TOUCH the inside of the Petri dish or expose it to outside air until you are ready to inoculate.**

- Your group will inoculate multiple plates for each dilution. The final plate will be inoculated with sterile dilution water only (**blank**), as a contamination control.

- Using a P200 pipet with sterile tip, transfer 0.1 mL of each dilution to each appropriately labeled Petri dish. Start with the most dilute concentration (1/10,000) and work your way up to the most concentrated (1/100).

- Using a tilting motion, gently spread the 0.1 mL of sample uniformly across the surface of the agar.

- Then use a sterile L-spreading tool to distribute the liquid uniformly on the plate. Go around at least five times, so that the liquid is completely dispersed and absorbed by the agar. *Do not puncture the agar.*

6. **Incubate plates.** After inoculation of your spread plates, turn them upside down (inverted), and tape your labeled Petri dishes together with masking tape. The inoculated plates are then placed **inverted** in the incubator.

- Remember to record the temperature in the incubator to the nearest degree Celsius, using a thermometer. A beaker of water will be kept in the incubator to keep the humidity level high so that the plates do not dry out. Make sure the beaker has sufficient water in it before closing the door. Just like an oven, do not leave the door open for any length of time to conserve the temperature inside.

6. **Compute growth rates.** After 48 hours, remove your plates from the incubator and select a relatively isolated typical colony on each of your plates. Mark this colony so you can find it the next day (see Figure 22).

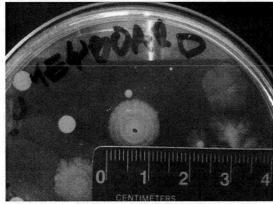

Figure 22. This Petri dish shows a typical colony marked with a dot, with a diameter of 1.2 cm
Source: Daniel Meeroff

- Using a ruler, record the diameter of your marked colonies starting after 24 hours, and each subsequent day for up to seven days. Using this data set, you can estimate a growth rate for each marked colony and record these values in a table in your report. **Note: Take digital photographs with a ruler so that you can compare colony sizes later on.**

7. **Count colony numbers and types.** After 48 hours incubation, take digital photographs of all spread plates (see Figure 23).

- Record the number and types of colonies on each of your spread plates in a table. **Note: Plates closest to the 20–300 colony range will give the most statistically accurate count.**

- For each plate, record morphological observations such as color, texture, colony diameter, shape, etc. in your table (see example spread plate data tables in the appendix).

- Do not forget to report the CFU/mL values in your table. *Remember, spread plate counts for HPC are made at 48 ± 4 hours incubation.*

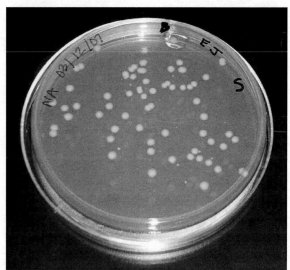

Figure 23. This Petri dish was made with 0.1 mL of 1/100 dilution and has 69 creamy white circle colonies and 57 tiny white colonies that are more difficult to see; therefore, the HPC = 126,000 CFU/mL is derived from two colony counts: 1) 69,000 CFU/mL for the creamy white circles and 2) 57,000 CFU/mL for the tiny white colonies
Source: Daniel Meeroff

8. **IDEXX Quanti-Tray*/2000 method.** An alternate way to conduct the HPC test is to use an IDEXX Quanti-Tray*/2000 method (www.idexx.com).

- Using the 1/100 dilution, pipet 1.0 mL into a sterile 100 mL vessel and fill to the mark with sterile dilution water. Do NOT overfill.

- Add contents of one media pack to the vessel, cap and shake to mix powder thoroughly.

Courtesy of Florida Atlantic University

- With one hand, hold the Quanti-Tray*/2000 upright with the well side facing you and squeeze the upper part so it bends toward you. Gently pull the foil tab to create a pocket. Avoid touching the inside of the sterile tray.

Courtesy of Florida Atlantic University

- Pour sample into a Quanti-Tray*/2000. Tap the small wells to release air bubbles and allow foam to settle.

- Then seal using a Quanti-tray sealer with orange rubber insert with well side facing down.

Courtesy of Florida Atlantic University

- Label the back of the tray with your team name, the sample identification, the date, and the dilution.

- Incubate at 35°C for 44 hours. Results are valid for up to 72 hours.

- Count the number of large and small blue fluorescent wells in the Quanti-Tray*/2000 using a 365-nm UV light in a dark environment.

Courtesy of Florida Atlantic University

- Interpret results using IDEXX Quanti-Tray*/2000 MPN Table in the appendix.

Discussion Questions

1. What does it mean if a sterile dilution water spread plate has growth on it? Did any of your sterile dilution water (**blank**) plates show any growth?

2. Provide a table of all of your heterotrophic plate counts and IDEXX MPN results for your natural water samples (see appendix for examples). Include the following data in your table: sample identification, colony types per plate, colony counts per plate, dilution factors, Heterotrophic Plate Count (CFU/mL), number of positive wells for, Heterotrophic Plate Counts for IDEXX (MPN/mL), averages of replicates, standard deviations, and relative errors for all dilutions and replicates. Compare your average HPC with the value obtained from the IDEXX HPC. Discuss the advantages and disadvantages of each and select the best option. Be sure to justify your answer with results.

3. Take all of your spread plate data with between 20 and 300 countable colonies per plate, and estimate the count on each of those using the quadrant method. Provide a table with the actual full colony count, the estimated colony count using the quadrant method, and the relative error between the two methods (see appendix).

- Show a sample calculation of how you estimated the value on these plates with the quadrant method.

- Did you have any plates that were not countable? Discuss why this happens.

4. Assume each bacterial cell occupies an area of 1.0 μm^2 and none of the cells are overlapping or stacked on top of one another. Now select a relatively isolated colony on two of your plates. Monitor these two colonies each day for up to seven days. Provide a table that shows the measured diameters (in mm) of each of these colonies each day (see appendix).

- Now, using these diameter values, compute the estimated number of bacteria cells in the colony on each day.

- Use the number of cells from two consecutive days (i.e., day 1 and day 2) to compute the exponential growth rate in $days^{-1}$ for each colony.

- Now estimate the colony size on day 7 based on the calculated exponential growth rates.

- Compare the estimated value with the actual number of cells in the colony on day 7 and report your relative error value. Comments?

Proper Safety and Disposal

Do NOT throw your Petri dishes in the trash!

All Petri dishes must be disposed of in the red bag within 7 days or the odor will become overwhelming. Viable bacteria water samples are not to be poured into the sink or placed directly in trash receptacles. They must be decontaminated first.

In case of accidental spill/exposure, hands and lower arms must be washed thoroughly with a germicidal soap. The spill kit contains the following items: Germicidal soap, Clorox wipes, 5% bleach solution, disposable gloves, and absorbent towels.

IDEXX trays are not required to be disinfected but must be placed in the red bag and disposed of in the biowaste collection container.

IDEXX Quanti-Tray®/2000 MPN Table

# Large Wells Positive	# Small Wells Positive																								
	0	1	2	3	4	5	6	7	8	9	10	11	12	13	14	15	16	17	18	19	20	21	22	23	24
0	<1	1.0	2.0	3.0	4.0	5.0	6.0	7.0	8.0	9.0	10.0	11.0	12.0	13.0	14.1	15.1	16.1	17.1	18.1	19.1	20.2	21.2	22.2	23.3	24.3
1	1.0	2.0	3.0	4.0	5.0	6.0	7.1	8.1	9.1	10.1	11.1	12.1	13.2	14.2	15.2	16.2	17.3	18.3	19.3	20.4	21.4	22.5	23.5	24.5	25.6
2	2.0	3.0	4.1	5.1	6.1	7.1	8.1	9.2	10.2	11.2	12.2	13.3	14.3	15.4	16.4	17.4	18.5	19.5	20.6	21.6	22.7	23.7	24.8	25.8	26.9
3	3.1	4.1	5.1	6.1	7.2	8.2	9.2	10.3	11.3	12.4	13.4	14.5	15.5	16.5	17.6	18.6	19.7	20.8	21.8	22.9	23.9	25.0	26.1	27.1	28.2
4	4.1	5.2	6.2	7.2	8.3	9.3	10.4	11.4	12.5	13.5	14.6	15.6	16.7	17.8	18.8	19.9	21.0	22.0	23.1	24.2	25.3	26.3	27.4	28.5	29.6
5	5.2	6.3	7.3	8.4	9.4	10.5	11.5	12.6	13.7	14.7	15.8	16.9	17.9	19.0	20.1	21.2	22.2	23.3	24.4	25.5	26.6	27.7	28.8	29.9	31.0
6	6.3	7.4	8.4	9.5	10.6	11.6	12.7	13.8	14.9	16.0	17.0	18.1	19.2	20.3	21.4	22.5	23.6	24.7	25.8	26.9	28.0	29.1	30.2	31.3	32.4
7	7.5	8.5	9.6	10.7	11.8	12.8	13.9	15.0	16.1	17.2	18.3	19.4	20.5	21.6	22.7	23.8	24.9	26.0	27.1	28.3	29.4	30.5	31.6	32.8	33.9
8	8.6	9.7	10.8	11.9	13.0	14.1	15.2	16.3	17.4	18.5	19.6	20.7	21.8	22.9	24.1	25.2	26.3	27.4	28.6	29.7	30.8	32.0	33.1	34.3	35.4
9	9.8	10.9	12.0	13.1	14.2	15.3	16.4	17.6	18.7	19.8	20.9	22.0	23.2	24.3	25.4	26.6	27.7	28.9	30.0	31.2	32.3	33.5	34.6	35.8	37.0
10	11.0	12.1	13.2	14.4	15.5	16.6	17.7	18.9	20.0	21.1	22.3	23.4	24.6	25.7	26.9	28.0	29.2	30.3	31.5	32.7	33.8	35.0	36.2	37.4	38.6
11	12.2	13.4	14.5	15.6	16.8	17.9	19.1	20.2	21.4	22.5	23.7	24.8	26.0	27.2	28.3	29.5	30.7	31.9	33.0	34.2	35.4	36.6	37.8	39.0	40.2
12	13.5	14.6	15.8	16.9	18.1	19.3	20.4	21.6	22.8	23.9	25.1	26.3	27.5	28.6	29.8	31.0	32.2	33.4	34.6	35.8	37.0	38.2	39.5	40.7	41.9
13	14.8	16.0	17.1	18.3	19.5	20.6	21.8	23.0	24.2	25.4	26.6	27.8	29.0	30.2	31.4	32.6	33.8	35.0	36.2	37.5	38.7	39.9	41.2	42.4	43.6
14	16.1	17.3	18.5	19.7	20.9	22.1	23.3	24.5	25.7	26.9	28.1	29.3	30.5	31.7	33.0	34.2	35.4	36.7	37.9	39.1	40.4	41.6	42.9	44.2	45.4
15	17.5	18.7	19.9	21.1	22.3	23.5	24.7	25.9	27.2	28.4	29.6	30.9	32.1	33.3	34.6	35.8	37.1	38.4	39.6	40.9	42.2	43.4	44.7	46.0	47.3
16	18.9	20.1	21.3	22.6	23.8	25.0	26.2	27.5	28.7	30.0	31.2	32.5	33.7	35.0	36.3	37.5	38.8	40.1	41.4	42.7	44.0	45.3	46.6	47.9	49.2
17	20.3	21.6	22.8	24.1	25.3	26.6	27.8	29.1	30.3	31.6	32.9	34.1	35.4	36.7	38.0	39.3	40.6	41.9	43.2	44.5	45.9	47.2	48.5	49.8	51.2
18	21.8	23.1	24.3	25.6	26.9	28.1	29.4	30.7	32.0	33.3	34.6	35.9	37.2	38.5	39.8	41.1	42.4	43.8	45.1	46.5	47.8	49.2	50.5	51.9	53.2
19	23.3	24.6	25.9	27.2	28.5	29.8	31.1	32.4	33.7	35.0	36.3	37.6	39.0	40.3	41.6	43.0	44.3	45.7	47.1	48.4	49.8	51.2	52.6	54.0	55.4
20	24.9	26.2	27.5	28.8	30.1	31.5	32.8	34.1	35.4	36.8	38.1	39.5	40.8	42.2	43.6	44.9	46.3	47.7	49.1	50.5	51.9	53.3	54.7	56.1	57.6
21	26.5	27.9	29.2	30.5	31.8	33.2	34.5	35.9	37.3	38.6	40.0	41.4	42.8	44.1	45.5	46.9	48.4	49.8	51.2	52.6	54.1	55.5	56.9	58.4	59.9
22	28.2	29.5	30.9	32.3	33.6	35.0	36.4	37.7	39.1	40.5	41.9	43.3	44.8	46.2	47.6	49.0	50.5	51.9	53.4	54.8	56.3	57.8	59.3	60.8	62.3
23	29.9	31.3	32.7	34.1	35.5	36.8	38.3	39.7	41.1	42.5	43.9	45.4	46.8	48.3	49.7	51.2	52.7	54.2	55.6	57.1	58.6	60.2	61.7	63.2	64.7
24	31.7	33.1	34.5	35.9	37.3	38.8	40.2	41.7	43.1	44.6	46.0	47.5	49.0	50.5	52.0	53.5	55.0	56.5	58.0	59.5	61.1	62.6	64.2	65.8	67.3
25	33.6	35.0	36.4	37.9	39.3	40.8	42.2	43.7	45.2	46.7	48.2	49.7	51.2	52.7	54.3	55.8	57.3	58.9	60.5	62.0	63.6	65.2	66.8	68.4	70.0
26	35.5	36.9	38.4	39.9	41.4	42.8	44.3	45.9	47.4	48.9	50.4	52.0	53.5	55.1	56.7	58.2	59.8	61.4	63.0	64.7	66.3	67.9	69.6	71.2	72.9
27	37.4	38.9	40.4	42.0	43.5	45.0	46.5	48.1	49.6	51.2	52.8	54.4	56.0	57.6	59.2	60.8	62.4	64.1	65.7	67.4	69.1	70.8	72.5	74.2	75.9
28	39.5	41.0	42.6	44.1	45.7	47.3	48.8	50.4	52.0	53.6	55.2	56.9	58.5	60.2	61.8	63.5	65.2	66.9	68.6	70.3	72.0	73.7	75.5	77.3	79.0
29	41.7	43.2	44.8	46.4	48.0	49.6	51.2	52.8	54.5	56.1	57.8	59.5	61.2	62.9	64.6	66.3	68.0	69.8	71.5	73.3	75.1	76.9	78.7	80.5	82.4
30	43.9	45.5	47.1	48.7	50.4	52.0	53.7	55.4	57.1	58.8	60.5	62.2	63.9	65.7	67.5	69.3	71.0	72.9	74.7	76.5	78.3	80.2	82.1	84.0	85.9
31	46.2	47.9	49.5	51.2	52.9	54.6	56.3	58.0	59.8	61.6	63.3	65.1	66.9	68.7	70.5	72.4	74.2	76.1	78.0	79.9	81.8	83.7	85.7	87.6	89.6
32	48.7	50.4	52.1	53.8	55.6	57.3	59.1	60.9	62.7	64.5	66.3	68.2	70.0	71.9	73.8	75.7	77.6	79.5	81.5	83.5	85.4	87.5	89.5	91.5	93.6
33	51.2	53.0	54.8	56.5	58.3	60.2	62.0	63.8	65.7	67.6	69.5	71.4	73.3	75.2	77.2	79.2	81.2	83.2	85.2	87.3	89.3	91.4	93.6	95.7	97.8
34	53.9	55.7	57.6	59.4	61.3	63.1	65.0	67.0	68.9	70.8	72.8	74.8	76.8	78.8	80.8	82.9	84.9	87.1	89.2	91.4	93.5	95.7	97.9	100.2	102.4
35	56.8	58.6	60.5	62.4	64.4	66.3	68.3	70.3	72.3	74.3	76.3	78.4	80.5	82.6	84.7	86.9	89.0	91.3	93.5	95.7	98.0	100.3	102.6	105.0	107.3
36	59.8	61.7	63.7	65.7	67.7	69.7	71.7	73.8	75.9	78.0	80.1	82.3	84.5	86.7	88.9	91.2	93.5	95.8	98.1	100.5	102.9	105.3	107.7	110.2	112.7
37	62.9	65.0	67.0	69.1	71.2	73.3	75.4	77.6	79.8	82.0	84.2	86.5	88.8	91.1	93.4	95.8	98.2	100.6	103.1	105.7	108.1	110.7	113.2	115.9	118.6
38	66.3	68.4	70.6	72.7	74.9	77.1	79.3	81.6	83.9	86.2	88.6	91.0	93.4	95.8	98.3	100.8	103.4	105.9	108.6	111.2	113.9	116.6	119.4	122.2	125.0
39	70.0	72.2	74.4	76.7	78.9	81.3	83.6	86.0	88.4	90.9	93.4	95.9	98.4	101.0	103.6	106.2	108.9	111.6	114.6	117.4	120.3	123.2	126.1	129.2	132.2
40	73.8	76.2	78.5	80.9	83.3	85.7	88.2	90.8	93.3	95.9	98.5	101.2	103.9	106.5	109.5	112.4	115.3	118.2	121.2	124.3	127.4	130.5	133.7	137.0	140.3
41	78.0	80.5	82.9	85.5	88.0	90.6	93.3	95.9	98.7	101.4	104.3	107.1	110.0	113.0	116.0	119.1	122.2	125.4	128.7	132.0	135.4	138.8	142.3	145.9	149.5
42	82.6	85.2	87.8	90.5	93.2	96.0	98.8	101.7	104.6	107.6	110.6	113.7	116.9	120.1	123.4	126.7	130.1	133.6	137.2	140.8	144.5	148.3	152.2	156.1	160.2
43	87.6	90.4	93.2	96.0	99.0	101.9	105.0	108.1	111.2	114.5	117.8	121.1	124.6	128.1	131.7	135.4	139.1	143.0	147.0	151.0	155.2	159.4	163.8	168.2	172.8
44	93.1	96.1	99.1	102.2	105.4	108.6	111.9	115.3	118.7	122.3	125.9	129.6	133.4	137.4	141.4	145.5	149.7	154.1	158.5	163.1	167.9	172.7	177.7	182.9	188.2
45	99.3	102.5	105.8	109.2	112.6	116.2	119.8	123.6	127.4	131.4	135.4	139.6	143.9	148.3	152.9	157.6	162.4	167.4	172.6	178.0	183.5	189.2	195.1	201.2	207.5
46	106.3	109.8	113.4	117.2	121.0	125.0	129.1	133.3	137.6	142.1	146.7	151.5	156.5	161.6	167.0	172.5	178.2	184.2	190.4	196.8	203.5	210.5	217.8	225.4	233.3
47	114.3	118.3	122.4	126.6	130.9	135.4	140.1	145.0	150.0	155.3	160.7	166.4	172.3	178.5	184.9	191.8	198.9	206.4	214.2	222.4	231.0	240.0	249.6	259.5	270.0
48	123.9	128.4	133.1	137.9	143.0	148.3	153.9	159.7	165.8	172.2	178.9	186.0	193.5	201.4	209.8	218.7	228.2	238.2	248.9	260.3	272.3	285.1	298.7	313.0	328.2
49	135.5	140.8	146.4	152.3	158.5	165.0	172.0	179.3	187.2	195.6	204.6	214.3	224.7	235.9	248.1	261.3	275.5	290.9	307.6	325.5	344.8	365.4	387.3	410.6	435.2

IDEXX Quanti-Tray®/2000 MPN Table

# Large Wells Positive	# Small Wells Positive																							
	25	26	27	28	29	30	31	32	33	34	35	36	37	38	39	40	41	42	43	44	45	46	47	48
0	25.3	26.4	27.4	28.4	29.5	30.5	31.5	32.6	33.6	34.7	35.7	36.8	37.8	38.9	40.0	41.0	42.1	43.1	44.2	45.3	46.3	47.4	48.5	49.5
1	26.6	27.7	28.7	29.8	30.8	31.9	32.9	34.0	35.0	36.1	37.2	38.2	39.3	40.4	41.4	42.5	43.6	44.7	45.7	46.8	47.9	49.0	50.1	51.2
2	27.9	29.0	30.0	31.1	32.2	33.2	34.3	35.4	36.5	37.5	38.6	39.7	40.8	41.9	43.0	44.0	45.1	46.2	47.3	48.4	49.5	50.6	51.7	52.8
3	29.3	30.4	31.4	32.5	33.6	34.7	35.8	36.8	37.9	39.0	40.1	41.2	42.3	43.4	44.5	45.6	46.7	47.8	48.9	50.0	51.2	52.3	53.4	54.5
4	30.7	31.8	32.8	33.9	35.0	36.1	37.2	38.3	39.4	40.5	41.6	42.8	43.9	45.0	46.1	47.2	48.3	49.5	50.6	51.7	52.9	54.0	55.1	56.3
5	32.1	33.2	34.3	35.4	36.5	37.6	38.7	39.9	41.0	42.1	43.2	44.4	45.5	46.6	47.7	48.9	50.0	51.2	52.3	53.5	54.6	55.8	56.9	58.1
6	33.5	34.7	35.8	36.9	38.0	39.2	40.3	41.4	42.6	43.7	44.8	46.0	47.1	48.3	49.4	50.6	51.7	52.9	54.1	55.2	56.4	57.6	58.7	59.9
7	35.0	36.2	37.3	38.4	39.6	40.7	41.9	43.0	44.2	45.3	46.5	47.7	48.8	50.0	51.2	52.3	53.5	54.7	55.9	57.1	58.3	59.4	60.6	61.8
8	36.6	37.7	38.9	40.0	41.2	42.3	43.5	44.7	45.9	47.0	48.2	49.4	50.6	51.8	53.0	54.1	55.3	56.5	57.7	59.0	60.2	61.4	62.6	63.8
9	38.1	39.3	40.5	41.6	42.8	44.0	45.2	46.4	47.6	48.8	50.0	51.2	52.4	53.6	54.8	56.0	57.2	58.4	59.7	60.9	62.1	63.4	64.6	65.8
10	39.7	40.9	42.1	43.3	44.5	45.7	46.9	48.1	49.3	50.6	51.8	53.0	54.2	55.5	56.7	57.9	59.2	60.4	61.7	62.9	64.2	65.4	66.7	67.9
11	41.4	42.6	43.8	45.0	46.3	47.5	48.7	49.9	51.2	52.4	53.7	54.9	56.1	57.4	58.6	59.9	61.2	62.4	63.7	65.0	66.3	67.5	68.8	70.1
12	43.1	44.3	45.6	46.8	48.1	49.3	50.6	51.8	53.1	54.3	55.6	56.8	58.1	59.4	60.7	62.0	63.2	64.5	65.8	67.1	68.4	69.7	71.0	72.4
13	44.9	46.1	47.4	48.6	49.9	51.2	52.5	53.7	55.0	56.3	57.6	58.9	60.2	61.5	62.8	64.1	65.4	66.7	68.0	69.3	70.7	72.0	73.3	74.7
14	46.7	48.0	49.3	50.5	51.8	53.1	54.4	55.7	57.0	58.3	59.6	60.9	62.3	63.6	64.9	66.3	67.6	68.9	70.3	71.6	73.0	74.4	75.7	77.1
15	48.6	49.9	51.2	52.5	53.8	55.1	56.4	57.8	59.1	60.4	61.8	63.1	64.5	65.8	67.2	68.5	69.9	71.3	72.6	74.0	75.4	76.8	78.2	79.6
16	50.5	51.8	53.2	54.5	55.8	57.2	58.5	59.9	61.2	62.6	64.0	65.3	66.7	68.1	69.5	70.9	72.3	73.7	75.1	76.5	77.9	79.3	80.8	82.2
17	52.5	53.9	55.2	56.6	58.0	59.3	60.7	62.1	63.5	64.9	66.3	67.7	69.1	70.5	71.9	73.3	74.8	76.2	77.6	79.1	80.5	82.0	83.5	84.9
18	54.6	56.0	57.4	58.8	60.2	61.6	63.0	64.4	65.8	67.2	68.6	70.1	71.5	73.0	74.4	75.9	77.3	78.8	80.3	81.8	83.3	84.8	86.3	87.8
19	56.8	58.2	59.6	61.0	62.4	63.9	65.3	66.8	68.2	69.7	71.1	72.6	74.1	75.5	77.0	78.5	80.0	81.5	83.1	84.6	86.1	87.6	89.2	90.7
20	59.0	60.4	61.9	63.3	64.8	66.3	67.7	69.2	70.7	72.2	73.7	75.2	76.7	78.2	79.8	81.3	82.8	84.4	85.9	87.5	89.1	90.7	92.2	93.8
21	61.3	62.8	64.3	65.8	67.3	68.8	70.3	71.8	73.3	74.9	76.4	77.9	79.5	81.1	82.6	84.2	85.8	87.4	89.0	90.6	92.2	93.8	95.4	97.1
22	63.8	65.3	66.8	68.3	69.8	71.4	72.9	74.5	76.1	77.6	79.2	80.8	82.4	84.0	85.6	87.2	88.9	90.5	92.1	93.8	95.5	97.1	98.8	100.5
23	66.3	67.8	69.4	71.0	72.5	74.1	75.7	77.3	78.9	80.5	82.2	83.8	85.4	87.1	88.7	90.4	92.1	93.8	95.5	97.2	98.9	100.6	102.4	104.1
24	68.9	70.5	72.1	73.7	75.3	77.0	78.6	80.3	81.9	83.6	85.2	86.9	88.6	90.3	92.0	93.8	95.5	97.2	99.0	100.7	102.5	104.3	106.1	107.9
25	71.7	73.3	75.0	76.6	78.3	80.0	81.7	83.3	85.1	86.8	88.5	90.2	92.0	93.7	95.5	97.3	99.1	100.9	102.7	104.5	106.3	108.2	110.0	111.9
26	74.6	76.3	78.0	79.7	81.4	83.1	84.8	86.6	88.4	90.1	91.9	93.7	95.6	97.4	99.2	101.0	102.9	104.7	106.6	108.5	110.4	112.3	114.2	116.2
27	77.6	79.4	81.1	82.9	84.6	86.4	88.2	90.0	91.9	93.7	95.5	97.4	99.3	101.2	103.1	105.0	106.9	108.8	110.8	112.7	114.7	116.7	118.7	120.7
28	80.8	82.6	84.4	86.3	88.1	89.9	91.8	93.7	95.6	97.5	99.4	101.3	103.3	105.2	107.2	109.2	111.2	113.2	115.2	117.3	119.3	121.4	123.5	125.6
29	84.2	86.1	87.9	89.8	91.7	93.7	95.6	97.5	99.5	101.5	103.5	105.5	107.5	109.6	111.6	113.7	115.7	117.8	119.9	122.1	124.2	126.4	128.6	130.8
30	87.8	89.7	91.7	93.6	95.6	97.6	99.6	101.6	103.7	105.7	107.8	109.9	112.0	114.2	116.3	118.5	120.6	122.8	125.1	127.3	129.5	131.8	134.1	136.4
31	91.6	93.6	95.6	97.7	99.7	101.8	103.9	106.0	108.2	110.3	112.5	114.7	116.9	119.1	121.4	123.6	125.9	128.2	130.5	132.9	135.3	137.7	140.1	142.5
32	95.7	97.8	99.9	102.0	104.2	106.3	108.5	110.7	113.0	115.2	117.5	119.8	122.1	124.5	126.8	129.2	131.6	134.0	136.5	139.0	141.5	144.0	146.6	149.1
33	100.0	102.2	104.4	106.6	108.9	111.2	113.5	115.8	118.2	120.5	122.9	125.4	127.8	130.3	132.9	135.3	137.8	140.4	143.0	145.6	148.3	150.9	153.7	156.4
34	104.7	107.0	109.3	111.7	114.0	116.4	118.9	121.3	123.8	126.3	128.8	131.4	134.0	136.6	139.2	141.9	144.6	147.4	150.1	152.9	155.7	158.6	161.5	164.4
35	109.7	112.2	114.6	117.1	119.6	122.2	124.7	127.3	129.9	132.6	135.3	138.0	140.8	143.6	146.4	149.2	152.1	155.0	158.0	161.0	164.0	167.1	170.2	173.3
36	115.2	117.8	120.4	123.0	125.7	128.4	131.1	133.9	136.7	139.5	142.4	145.3	148.3	151.3	154.3	157.3	160.5	163.6	166.8	170.0	173.3	176.6	179.9	183.3
37	121.3	124.0	126.8	129.6	132.4	135.3	138.2	141.2	144.2	147.3	150.3	153.5	156.7	159.9	163.1	166.4	169.8	173.2	176.7	180.2	183.7	187.3	191.0	194.7
38	127.9	130.8	133.8	136.8	139.9	143.0	146.2	149.4	152.6	155.9	159.2	162.6	166.1	169.6	173.2	176.8	180.4	184.2	188.0	191.8	195.7	199.7	203.7	207.7
39	135.3	138.5	141.7	145.0	148.3	151.7	155.1	158.6	162.1	165.7	169.4	173.1	176.9	180.7	184.7	188.7	192.7	196.8	201.0	205.3	209.6	214.0	218.5	223.0
40	143.7	147.1	150.6	154.2	157.8	161.5	165.3	169.1	173.0	177.0	181.0	185.2	189.4	193.7	198.1	202.5	207.1	211.7	216.4	221.1	226.0	231.0	236.0	241.1
41	153.2	157.0	160.9	164.8	168.9	173.0	177.2	181.5	185.8	190.3	194.8	199.5	204.2	209.1	214.0	219.1	224.2	229.4	234.8	240.2	245.8	251.5	257.2	263.1
42	164.3	168.6	172.9	177.3	181.9	186.5	191.3	196.1	201.1	206.2	211.4	216.7	222.2	227.7	233.4	239.2	245.2	251.3	257.5	263.8	270.3	276.9	283.6	290.5
43	177.5	182.3	187.3	192.4	197.6	202.9	208.4	214.0	219.8	225.8	231.8	238.1	244.5	251.0	257.7	264.6	271.7	278.9	286.3	293.8	301.5	309.4	317.4	325.7
44	193.6	199.3	205.1	211.0	217.2	223.5	230.0	236.7	243.6	250.8	258.1	265.6	273.3	281.2	289.4	297.8	306.3	315.1	324.1	333.3	342.8	352.4	362.3	372.4
45	214.1	220.9	227.9	235.3	242.7	250.4	258.4	266.7	275.3	284.1	293.3	302.6	312.3	322.3	332.5	343.0	353.8	364.9	376.2	387.9	399.8	412.0	424.5	437.4
46	241.5	250.0	258.9	268.2	277.8	287.8	298.1	308.8	319.9	331.4	343.3	355.5	368.1	381.1	394.5	408.3	422.5	437.1	452.0	467.4	483.3	499.6	516.3	533.5
47	280.9	292.4	304.4	316.9	330.0	343.6	357.8	372.5	387.7	403.4	419.8	436.6	454.1	472.1	490.7	509.8	529.8	550.4	571.7	593.8	616.7	640.5	665.3	691.0
48	344.1	360.9	378.4	396.8	416.0	436.0	456.9	478.6	501.2	524.7	549.3	574.8	601.5	629.4	658.6	689.3	721.5	755.6	791.5	829.7	870.4	913.9	960.6	1011.2
49	461.1	488.4	517.2	547.5	579.4	613.1	648.8	686.7	727.0	770.1	816.1	866.4	920.8	980.4	1046.9	1119.9	1203.3	1299.7	1413.6	1553.1	1732.9	1986.3	2419.6	>2419.6

Example Spread Plate Data Tables

Sample Identification	Colony Type	Numbers after 48 hours	mL of sample in Vial	mL of dilution water in vial	Dilution factor	mL on Petri Dish	Final dilution factor	HPC (CFU/mL)
Canal	White dots	35	0.1	9.9	1/100	0.1	1/1000	35,000
	Creamy circles	25	0.1	9.9	1/100	0.1	1/1000	25,000
	Orange dots	2	0.1	9.9	1/100	0.1	1/1000	2,000
	TOTAL	**62**	**0.1**	**9.9**	**1/100**	**0.1**	**1/1000**	**62,000**
Sea	White dots	52	0.1	9.9	1/100	0.1	1/1000	52,000
	Yellow circles	12	0.1	9.9	1/100	0.1	1/1000	12,000
	TOTAL	**64**	**0.1**	**9.9**	**1/100**	**0.1**	**1/1000**	**64,000**

Sample Identification	Number of positive large wells	Number of positive small wells	MPN value of positive wells	mL of sample in vial	mL of dilution water in vial	Dilution factor	Bottle dilution	IDEXX HPC (MPN/mL)	Average	Std. Dev.
Canal 1/100	0	6	6.0	0.1	9.9	1/100	1/100	60,000		
Canal 1/100	5	1	6.3	0.1	9.9	1/100	1/100	63,000		
	TOTAL								61,500	2121
Sea 1/100	3	2	5.1	0.1	9.9	1/100	1/100	51,000		
Sea 1/100	2	4	6.1	0.1	9.9	1/100	1/100	61,000		
	TOTAL								56,000	7071

Sample	HPC (CFU/mL)	IDEXX (MPN/mL)	Percent Error
Canal	62,000	60,000	3.3%
		63,000	-1.6%
Sea	64,000	51,000	23%
		61,000	4.8%

Sample Identification	Numbers after 48 hours	Quadrant 1	Quadrant 2	Quadrant 3	Quadrant 4	Average	HPC	Estimated HPC	Percent Error
Canal 1/100A	271	81	75	61	54	68	27,100	27,200	0.4%
Canal 1/100B	247	201	12	21	13	62	20,100	24,800	20.9%
Sea 1/100A	211	74	39	47	51	53	21,100	21,200	0.5%
Sea 1/100B	258	55	84	73	46	64	25,800	25,600	0.8%
Sea 1/1000A	35	6	8	7	14	9	35,000	36,000	2.8%
Sea 1/1000B	22	20	2	0	0	6	22,000	24,000	8.7%

Sample	Day	Size	N_t		r (day^{-1})	Average r (day^{-1})	Predicted using average r	Error
1/100 Canal	1	8	5.02E+07	R1→2	0.64	0.44		
	2	11	9.50E+07	R2→5	0.29			
	5	14	2.27E+08	R1→5	0.38			
	7	29	6.60E+08				8.89E+08	34.7%

$$\text{Day 7 error} = \frac{\left| N_7 - N_{7\,\text{Predicted}} \right|}{N_7} = \frac{\left| 6.60\text{E}+08 - 8.89\text{E}+08 \right|}{6.60\text{E}+08} \times 100\% = 34.7\%$$

Activity #4. Disinfection

Purpose:

- To determine the effect of disinfecting agents on microbial growth patterns.

Background

Microorganisms are present just about everywhere (*ubiquitous*) in the environment, on surfaces, in the soil, suspended in the air, and even inside your body. Since you will be working with environmental samples, there is a potential that pathogenic microorganisms might be present. Pathogens can cause disease in humans. Therefore, care must be taken to prevent possible infection or transmission from the laboratory.

Antiseptics and disinfectants are chemical substances that help prevent infection. **Disinfectants** are anti-microbial substances that inactivate or otherwise limit the growth of microorganisms, such as bacteria and fungi. The most effective agents are highly germicidal, having a rapid effect on a wide variety of microorganisms (broad-spectrum). They should also be relatively inexpensive and aesthetically acceptable. In addition, the ingredients should have low toxicity to humans, so that they can be applied safely.

Microorganisms vary in their resistance to disinfection by physical or chemical means. A disinfectant that inactivates bacteria may be ineffective against viruses or fungi. There are differences in susceptibility even between strains of the same species. Bacterial spores and protozoan cysts and oocysts are more resistant than vegetative forms, and non-enveloped, non-protein coat-containing viruses respond differently than do other viruses.

You are probably familiar with the most common types of disinfection agents, including high temperatures (autoclaving, boiling, etc.), physical agents (ultraviolet radiation, membrane filtration, etc.), and chemical agents (bleach, iodine, alcohol, hydrogen peroxide, quaternary ammonium, phenols, etc.). The mode of action varies. Some agents disrupt DNA. Others burst the cell open, and others interrupt essential cell functions.

Handwashing is a simple and effective way to limit the transmission of communicable diseases. While antibacterial soap may provide some additional protection, the major effect of handwashing is the mechanical removal of microbes from the skin. Friction when washing hands is important to mechanically remove microorganisms from the surface of the skin. Using a paper towel to turn off the water prevents recontamination of the hands with microorganisms.

Pre-Lab Questions

1. Make a table that shows the four samples you selected for the disinfection procedure, arrange the values from highest to lowest expected HPC.

2. Describe the mode of action of the disinfection agents that you have selected for this experiment.

Procedure

1. **Prepare Swab Testing Plates.** Each group will be furnished with several Petri dishes for swab testing.
- The following swabs will be collected:

   1) Initial surface swab

2) Surface swab washed with a water rinse

3) Surface swab washed with an anti-bacterial agent

4) Sterile dilution water control (**blank**)

5) Any other surface as directed by your instructor

2. **Collect Surface Samples.** Remove the sterile swab from its package aseptically, by tearing open the package without touching the sterile tip of the swab.

- Moisten the swab tip in sterile dilution water first, before swabbing the surface.

- Then using a rolling motion, gently swab the desired surface thoroughly (see Figure 24).

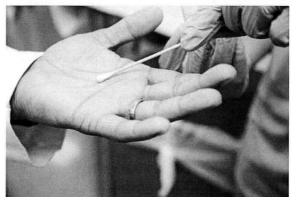

Figure 24. Swabbing technique for handwashing experiment
Courtesy of Florida Atlantic University

- Using a new sterile swab, collect the second sample after rinsing the surface with tap water *without soap*, and shaking off the excess water. **Do not dry with a towel**.

- Using a new sterile swab, collect the third sample by washing and scrubbing thoroughly with an anti-bacterial agent, shaking off the excess water. **Again, do not dry with a towel**.

- For the sterile dilution water control sample (**blank**), moisten a new sterile swab tip in the sterile dilution water provided.

3. **Prepare Samples for Plate Inoculation.**
- Dip the swab tip into a separate labeled vial containing 1.0 mL of sterile dilution water.

- Break off the excess wooden tip so that the swab end is completely underwater and the cap will close.

- Close the vial cap and swirl vigorously to transfer the microorganisms from the swab to the water in the vial. Do not invert since the vial cap is not sterile.

- Using a P200 pipet with sterile tip, transfer 0.1 mL to the appropriately labeled Petri dish. Use a new sterile pipet tip for each sample and discard the tip after inoculation.

- Using a tilting motion gently spread the 0.1 mL drop of sample uniformly across the surface of the agar. Then use a sterile L-spreading tool to distribute the liquid uniformly on the plate. Go around five times, so that the liquid is completely dispersed and absorbed by the agar. *Do not puncture the agar*.

4. **Incubate Plates.** After inoculation, tape your labeled Petri dishes together with masking tape. The inoculated plates will be incubated **inverted** in a warm, dark environment for <u>48 hours</u>.

- Remember to record the temperature in the incubator to the nearest degree Celsius, using a thermometer. A beaker of water will be kept in the incubator to keep the humidity level high so that the plates do not dry out.

5. **Count Colony Numbers and Types**. Take digital photographs of all your plates, daily.

- Using a ruler, record the diameter of your marked colonies starting after 24 hours, and each subsequent day for up to 7 days. Using this data set, you can estimate a growth rate for each marked colony and record these values in a table in your report. **Note: Take digital photographs with a ruler so that you can compare colony sizes later on.**

- Heterotrophic Plate Counts will be made after 48 ± 4 hours have elapsed. Record the number and types of colonies on your spread plates, and report the CFU/mL values in a table.

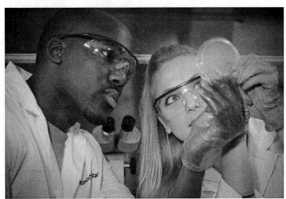

Courtesy of Florida Atlantic University

Discussion Questions

1. What are your estimated microbial concentration values in each of your samples, and how do these values compare to typical values reported in the literature?

2. Provide a table of all of your spread plates from the swab tests, <u>arranged in order</u> from highest diversity to lowest diversity. If a monoculture is observed (only one type of colony), does this mean that no other microorganisms are present? Why?

3. (2 pointer) Provide one bar graph and brief discussion of results as to which of the hand washing techniques shows the highest HPC. Now rearrange the order of observations to correspond to the order of swabbing (for the hand washing experiment). Did the order of swabbing your plates have any observable effect on the results? Explain.

Proper Safety and Disposal

Do NOT throw Petri dishes in the trash!

All Petri dishes must be disposed of in the biohazardous waste red bag within 7 days. Water samples are not to be poured into the sink or placed directly in trash receptacles. They must be decontaminated first.

In case of accidental spill/exposure, hands and lower arms must be washed thoroughly with a germicidal soap. The spill kit contains the following items: Germicidal soap, Clorox wipes, 5% bleach solution, vinyl or latex gloves, and absorbent towels.

Appendix

Example Swab Data Set

Sample	Colony Type 1	Colony Type 2	Colony Type 3	Lawn Present?	Total	Diversity	Dilution Factor	CFU/mL	Reported Value
Unwashed Hand	100	50	--	No	150	2	10	1500	1500
Water Rinse	80	20	--	No	100	2	10	1000	1000
Spice Foam	20	10	5	No	35	3	10	350	350
Sterile Water	2	--	--	No	2	1	10	20	<20
Purell	--	--	--	No	0	0	10	0	<20

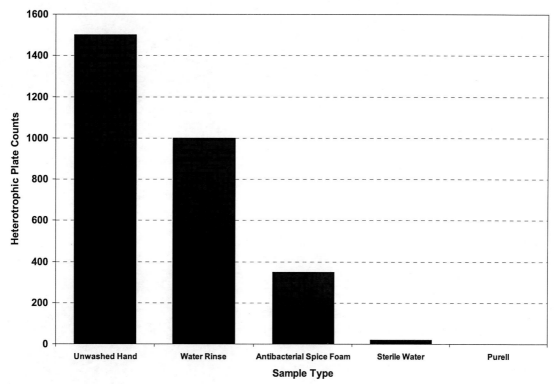

Source: Daniel Meeroff

Activity #5. Oxygen Demand

Purpose:

- To determine the amount of biodegradable and non-biodegradable organic matter in a natural water sample.

Background

Because of the enormous variety of organic compounds that exist in natural waters, determining the exact content is a complex endeavor. Therefore, scientists and engineers have come up with a surrogate measure of water pollution. This method relies on the fact that when organics decompose, an equivalent amount of dissolved oxygen (DO) is consumed. The total amount of organics is determined by a chemical reaction using the COD test, and the biodegradable fraction is determined by a microbial reaction using the BOD test.

The **chemical oxygen demand (COD)** measures the oxygen equivalent of the organic content that is oxidized by a strong chemical oxidant. Thus, it includes *both* the biodegradable component and the refractory (non-biodegradable) component.

The COD test is relatively fast, taking just 2 to 3 hours, as opposed to the BOD_5 test, which requires 5 days to complete. Another difference is that BOD is a biochemical process as measured by the ability of microbes to degrade the organics, while COD is purely a chemical oxidation process.

To analyze for COD, we use the dichromate reflux method by boiling in a mixture of potassium dichromate ($K_2Cr_2O_7$), concentrated sulfuric acid (H_2SO_4), and a mercury salt. Digestion occurs by heating the samples in an oven or block digester at 150°C (Figure 25).

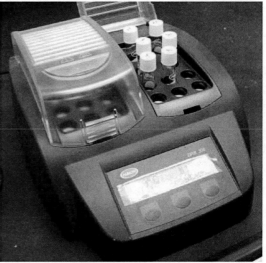

Figure 25. COD digestion heating block
Source: Daniel Meeroff

After two hours of reflux (boiling in a sealed container), the tubes are removed from the digester, cooled, and measured spectrophotometrically at λ = 620-nm (Figure 26).

Figure 26. Spectrophotometer for measuring COD
Source: Daniel Meeroff

When a sample is digested, the organic material in the sample gets oxidized and reduces the orange dichromate ion

($Cr_2O_7^{2-}$) to the green chromic ion (Cr^{3+}), which absorbs strongly at 620-nm, where the dichromate ion has nearly zero absorption. Thus after digestion, the remaining unreduced dichromate is related to the oxidizable organic matter in terms of oxygen equivalent.

Results from the COD test will be used to calculate the required dilutions for the BOD test.

The biochemical oxygen demand (BOD) is the amount of oxygen required by microorganisms to biologically degrade organic wastes. Typically, the BOD test is used to measure the strength of organic pollution. In fact, BOD is commonly used as an indicator to:

1. Determine compliance with wastewater discharge permits

2. Monitor wastewater treatment plant operation and performance

3. Design wastewater treatment and disposal facilities

4. Estimate the assimilative capacity of the receiving body of water.

Measuring the amount of organics in a sample requires complete stabilization of a waste by microorganisms (**ultimate BOD**). However, this requires an incubation period that is much too long (i.e. 60–90 days) for practical purposes.

Therefore, the 5-day test (BOD5) has been adopted as the standard. BOD5 is the total amount of oxygen consumed by microorganisms during the first 5 days of biodegradation. Samples are incubated at 20°C in the dark. This prevents algae from adding oxygen to the air-tight bottle. The larger the BOD value, the more biodegradable organics the sample contains. For instance, typical untreated domestic (raw) wastewater has a BOD5 concentration on the order of 100–300 mg/L (Metcalf & Eddy, Inc. 2003).

The organics measured during this 5-day period typically correspond to the carbonaceous BOD, since the bacteria that oxidize nitrogen are usually not present in sufficient numbers to influence oxygen consumption until approximately 7 days. However, it is common practice to use a nitrogen inhibitor to suppress nitrification.

Bacterial growth requires nutrients, including nitrogen, phosphorus, and trace elements. These nutrients are added to the dilution water, which is also buffered to ensure that the pH of the sample remains suitable for the bacteria present in the sample.

The BOD curve (Figure 27) is a first order expression:

$$\frac{dL_t}{dt} = -kL_t$$

Where:
L_t = O2 equivalent of organics remaining at time t (mg/L as O_2)
k = reaction rate constant ($time^{-1}$)

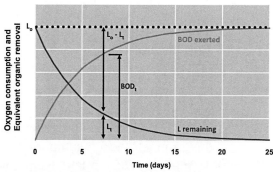

Figure 27. Example of a BOD curve
Source: Daniel Meeroff

Integrating the BOD curve yields:

$$y = BOD_t = L_o\left(1 - e^{-kt}\right)$$

Where:
  y = BOD at any time t (BOD_t, in mg/L as O_2)
  L_o = ultimate BOD (mg/L as O_2)

Evaluation of this equation is complicated because both L_o and k are generally unknown. However, if suitable data are available, procedures exist for determining both of these parameters, including the following:

1. First order method
2. Thomas slope method (Thomas 1937)
3. BOD_t vs. BOD_{t+1} method (Masters and Ela 2008)

First order method. First calculate the ln of your BOD data, then plot the $ln(BOD_t)$ versus t. The slope of the line is equal to the value of k. Then use this value and your last measured BOD_t value to compute L_o.

Thomas slope method. This approach is based on the following linear approximation:

$$\left(\frac{t}{y}\right)^{\frac{1}{3}} = \left(\frac{k^{2/3}}{6L_o^{1/3}}\right)t + (kL_o)^{-\frac{1}{3}}$$

A plot of $(t/y)^{1/3}$ versus t is a straight line with slope $m = k^{2/3}/(6L_o^{1/3})$ and intercept on the $(t/y)^{1/3}$ axis of $b = (kL_o)^{-1/3}$. It can then be shown that $k = 6m/b$ and then $L_o = 1/(kb^3)$. **Note:** Values of $y > 0.9L_o$ should NOT be used in this procedure because in this range the assumptions made in developing this relationship no longer apply.

BOD_t vs. BOD_{t+1} method. The third method is found in Masters and Ela (2008) Problem 5.14. All you have to do is plot BOD_t versus BOD_{t+1}. Then plot a 1:1 slope line, and from the point where these two lines intersect, read down to the x-axis, this is the L_o. You can solve for this value mathematically by taking the equation of the BOD_t vs. BOD_{t+1} line and setting it equal to x. Then solve for x, this will be your L_o. Now use this value and your last measured BOD_t value to compute k.

Dilutions and BOD_5 Calculations

The BOD of an unseeded water sample at any time t (BOD_t) is given by the following expression:

$$BOD_t = \frac{DO_{t=o} - DO_t}{P}$$

Where:
  $DO_{t=o}$ = Initial dissolved oxygen (mg/L as O_2)
  DO_t = Dissolved oxygen at time t (mg/L as O_2)
  P = volumetric fraction of sample used (P = mL sample / mL total)

For example, if 25 mL sample is used in a 300 mL bottle, then $P = 25 \div 300 = 0.083$.

Let's make sure you understand how to do this correctly.

Example 5.1. A sample has an initial dissolved oxygen (DO_0) of 9.0 mg/L assuming that the test does not require additional seed, and a standard BOD bottle contains 0.300 L, what is the BOD_5 if the 75 mL sample contains 4.9 mg/L as O_2 on the fifth day?

Solution:

- The DO deficit is given by:

$$(DO_{t=0} - DO_{t=5}) = 9.0 - 4.9 = \underline{4.1 \text{ mg/L as } O_2}$$

- The volumetric fraction P is given by:

$$P = 75 \div 300 = \underline{0.25}$$

- Finally, the BOD_5 is calculated as follows:

$$BOD_5 = 4.1 \div 0.25 = \underline{16 \text{ mg/L as } O_2}$$

Sometimes, the sample is too dilute, so we must seed with microorganisms to stimulate biodegradation. In this case, we must use two sets of sample bottles:

- Seeded blank
- Sample + seed

When using seed, the BOD calculation is given by:

$$BOD = \frac{(DO_{t=0} - DO_t) - (B_i - B_t)f}{P}$$

$$f = \frac{seed_{\text{diluted sample}}}{seed_{\text{control sample}}} = (1 - P)$$

Where:
B_t is the measured dissolved oxygen in the seeded dilution water at any time t.

The experimental procedure is based upon diluting a sample with a known amount of dilution water and seed and measuring the dissolved oxygen concentrations over the course of 5 days or more. One set of experiments is conducted on the dilution water and seed only, and another set is conducted on the mixture of dilution water and seed plus the environmental samples.

For a valid BOD test, the following conditions must apply:

- **Flag 1.** DO deficit: $(DO_{t=0} - DO_{t=5})$ must be larger than 2.0 mg/L
- **Flag 2.** Final DO ($DO_{t=5}$) must be larger than 1.0 mg/L
- **Flag 3.** Blank DO deficit ($DO_{t=0} - DO_{t=5}$) must be less than 0.2 mg/L
- **Flag 4.** Standard deviation of the samples must not exceed 30% of the mean

One way to estimate the required dilution (P) would be to take the COD value, apply a safety factor and use the conditions of validity of the BOD test (Flags 1 and 2).

Let's do an example to make sure that you understand how to estimate the dilution factor (P) for BOD from its COD value.

Example 5.2. Let's say the COD values for our sample average to 170 mg/L, and the BOD_5 is estimated to be 75% of the COD value. If the initial DO in the field was measured at 7.82 mg/L, and $P = x / 0.300$, then calculate P.

Solution:
Expected $BOD_5 = 0.75 COD = 0.75 \times 170$
$= \underline{130 \text{ mg/L as } O_2}$

$$\text{Flag 1}: 130 = \frac{2.0}{(x/0.300)} \rightarrow x = 0.00462 \, L$$

$$\text{Flag 2}: 130 = \frac{(7.82 - 1.0)}{(x/0.300)} \rightarrow x = 0.0157 \, L$$

Using Flag 1 (DO deficit > 2.0 mg/L), and Flag 2 ($DO_f > 1.0$ mg/L), would lead you to expect to use between 0.00462 – 0.0157 L of sample in your unseeded BOD dilution bottle. Perhaps you should select

0.010 L, which is in between the two extremes, so that your *P = 0.010/0.300 = 0.033*.

If one of your BOD samples fails to meet any of the flags, follow the example below to report your estimated BOD value:

Sample Volume	DO_o	DO_5	B_o	B_5	Reported BOD_5	Flag
0.150 L	9.81	8.18	6.85	6.15	< 3.30	1
0.150 L	9.81	0.75	6.85	6.15	> 16.9	2
0.150 L	9.81	4.95	6.85	6.15	9.02	--

In the first row of data, Flag 1 applies because the deficit is too low (9.81 − 8.18 = 1.63), so set the deficit to 2.0 and:

$$BOD_5 < \frac{(2.0) - (6.85 - 6.15)0.5}{0.5} < 3.30 \, mg/L$$

In the second row of data, Flag 2 applies because the DO_5 is less than 1.0 mg/L (0.75 < 1). Therefore, set the DO_5 value to 1.0 and:

$$BOD_5 > \frac{(9.81 - 1.0) - (6.85 - 6.15)0.5}{0.5} > 16.9 \, mg/L$$

You cannot report the BOD_5 unless it meets all of the validity criteria.

Pre-Lab Questions

1. Why is chemical oxygen demand important in engineering?

2. If your spectrophotometer reads 125 mg/L as O_2 and you diluted your COD sample by adding 0.500 mL of sample and 1.500 mL of dilution water, what is the COD in this sample in units of mg/L?

3. If your average BOD_5 value is 61 mg/L, your average k value is 0.19 day^{-1}, and your average COD value is 115 mg/L,

how much of the organic material is non-biodegradable, in percent?

4. Potassium acid phthalate, also known as KHP ($HOOCC_6H_4COOK$), is used to develop a standard curve for COD. Write the complete balanced equation for the oxidation of KHP by oxygen to form carbon dioxide, water, and potassium hydroxide. Calculate exactly how much O_2 (theoretical oxygen demand) is required to completely oxidize 5.000 grams of KHP.

5. What is your estimated ultimate BOD level in your samples? Provide a reference source for this value.

6. If you were running a BOD test without seeding, what dilution (*P*) should you use for this estimated BOD value?

Procedure

1. **Collect field samples.** Obtain 1-L samples of natural water, and filter with a coffee filter to remove any solids.

2. **Prepare COD samples.** Preheat the COD reactor to 150°C and turn on the spectrophotometer.

• Wearing protective gloves, remove the cap from the pre-dispensed reagent vial. Note: The reagent mixture is light sensitive. Label the cap with your sample identification code using a Sharpie.

• **Prepare Blank.** (Only one is needed per class). Holding the vial at a 45° angle, pipet 2.00 mL of dilution water into the vial. Replace the vial cap tightly, and rinse with deionized water, then wipe clean with a paper towel.

- **Prepare QA/QC Sample.** (Only one is needed per class). Repeat as above but use 2.00 mL of KHP standard solution.

- **Prepare Environmental Sample(s).** For most accurate results, repeat the analysis with a diluted sample.

  - **Diluted Sample.** In one vial, pipet 1.00 mL of sterile dilution water + 1.00 mL of filtered sample.

  - **Undiluted Samples.** In the other two vials, pipet 2.00 mL of filtered sample so that you have two replicates.

- Hold the vial by the cap over a sink and invert gently 20 times to mix the contents. Caution: Vials become hot after mixing.

- Wipe the glass vials with a wet towel, then wipe dry.

3. **Place the vials in the preheated COD reactor.** Begin the timer (2 hours). After 2 hours, allow the vials to come to room temperature in the dark.

4. **Analyze the samples.** Select the stored program number on the spectrophotometer. Make a note of the wavelength (λ) that is selected.

- **Analyze Blank:** Clean the outside of the vial with a Kimwipe. Note: Wiping with a damp towel, followed by a dry one, will remove fingerprints or other marks. Then place the blank into the adapter with the logo facing the front of the instrument. Close the light shield. Press the soft key under ZERO. The display will show: 0 mg/L COD.

- **Analyze QA/QC Sample:** To read samples, place the vial into the adapter with the logo facing the front of the instrument, close the light shield, and read the result on the screen. Record

results in terms of mg/L COD, then change the units to get the absorbance. Obtain the known COD value for this calibration check, and calculate the percent error.

- **Analyze Remaining Samples.** Record the COD results in terms of mg/L COD and absorbance units.

- **Analyze with Turbulence.** Take one of your previously analyzed sample vials, shake vigorously, and immediately read the COD value on the spectrophotometer to see what happens to the concentration.

- **Analyze with Different Wavelength.** Now change the wavelength on the instrument manually to 510-nm to see what happens when you use the incorrect wavelength. Measure the absorbance of your blank and QA/QC samples to compare to the values you recorded using the correct wavelength setting.

5. **Estimate BOD dilution factor.** Follow the procedure in Example 5.2 to estimate the amount of sample needed for the unseeded BOD bottle.

6. **Prepare BOD dilutions.** Prepare one dilution water blank, one seeded blank, and a series of sample dilutions. Add one (1) BOD powder pillow of nutrients to each sample bottle.

- **Dilution Water Blank:** Prepare one bottle using 300 mL of sterile dilution water, without seed.

- **Seeded Blank:** Prepare one seeded blank by using sterile dilution water spiked with a measured amount of seed (i.e., 1.0 – 2.0 mL seed per 300 mL), as shown in Figure 28.

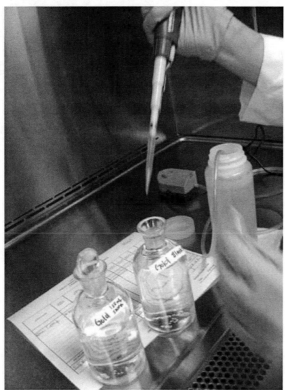

Figure 28. Adding seed to the BOD sample bottles
Source: Daniel Meeroff

- **Unseeded Sample**: Prepare one unseeded sample by using sterile dilution water spiked with a measured amount of sample as determined from the COD testing. *Note: Don't forget to mix your sample well, before making dilutions.*

- **Seeded Samples**: Prepare an assigned number of dilutions of each sample by adding the indicated amount of sample and/or seed (i.e., 250, 150, 100, 75, and 50 mL) to each BOD bottle. In each of the seeded sample bottles, add the same amount of seed as used in the seeded blank. Remember to label your team name on each bottle including blanks. *Note: Don't forget to mix your sample well, before making dilutions.*

7. **Analyze all samples.** Use a dissolved oxygen meter with BOD probe (Figure 29) to collect the data. Before making any measurements with the BOD probe for the day, make sure it is calibrated (with the probe in water-saturated air).

- On day zero, measure the initial DO of each sample bottle. Take the aluminum foil cap off the BOD bottle. Then remove the glass cap and place it inside the foil. Next rinse the BOD probe tip with deionized water and dry it with a Kimwipe before placing the tip in the sample bottle. Turn on the stirrer and check to make sure there are no air bubbles. *You may need to add a small amount of sterile water to eliminate the air bubbles.*

- Record the initial DO in mg/L as O_2 (**DO_o**). Also record the percent saturation, temperature in Celsius, the barometer reading, and the time of the reading. If the temperature is not 20°C, use the appendix chart to convert your DO values to 20°C. (see Example 5.3 in the appendix)

Figure 29. Dissolved oxygen meter for measuring BOD
Courtesy of Florida Atlantic University

- Top off all of your BOD bottles with sterile dilution water (if needed), and insert the glass stopper so that it seals, making sure no air bubbles are present. Then place the aluminum foil wrapping back on the cap to prevent evaporation.

- Place your sample bottles in the BOD incubator at exactly 20°C.

8. **Repeat DO measurements each day for up to 7 days.** Make sure to bring your raw data from the previous day every time.

- On day 1, measure the DO of each bottle. **This is DO$_1$.** Also record the temperature in Celsius, the percent saturation, the time, and the barometer reading.

- Each day, top off all of your BOD bottles with sterile dilution water (if needed), insert glass stoppers, and wrap with aluminum foil. Make sure if air bubbles are present that you make a note of it and try to eliminate them.

- Place your sample bottles back in the BOD incubator at exactly 20°C.

- Before you leave the lab, check to make sure your $DO_t > 1.0$ mg/L and the deficit is greater than 2.0 mg/L. If not, flag the data. Repeat measurements each day for up to 10 days or until $DO_f < 1.0$ mg/L.

9. **Plot all of your dilutions for BOD versus time, and determine if nitrification occurred.** Use the three methods described earlier to determine the BOD reaction constant k_1 and the ultimate BOD in mg/L as O_2 for each method.

10. **Check with your instructor before stopping data collection and properly disposing of samples.**

Discussion Questions

1. Why must the suspended material be allowed to settle before measuring absorbance with a spectrophotometer? Provide an example using one of your samples by listing the absorbance value before and after shaking.

2. Would the COD test still work if we measured the absorbance using a different wavelength? Provide a table that lists the absorbance value of the blank and one of your samples at one wavelength and at the original wavelength and also the percent error between the two measurements at different wavelengths.

Example Table:

Sample	Abs @ 610 nm	Abs @ 510 nm	% error
Blank	0.000	0.000	0.00%
5 :	0.022	-0.014	160%

3. Provide a table with your COD values for all of your samples and replicates in mg/L; include in the table the ultimate BOD values that you measured for those same samples, and determine the non-biodegradable fractions in each of your samples. Discuss.

4. Provide a table showing all of your BOD samples, and show if the results of each of the three critical criteria for a valid test were met or not. If any of your samples failed the critical criteria, provide a table listing what you could do to make each of them valid, if you had to repeat the test with the same sample.

5. Using your BOD$_t$ data table, plot BOD$_t$ versus time. Then determine k graphically by plotting $\ln(BOD)$ on the y-axis versus *time* on the x-axis. Using this k value and your final BOD$_t$, calculate ultimate BOD (L_o). This is the data for method 1. Now plot your data using the Thomas slope method to get k and L_o. This is the data for method 2. Now try computing the

ultimate BOD using the method described in Masters and Ela (2008). Use this value and the BOD_t value to compute k. This is the data for method 3. Now provide a table with all of your ultimate BOD results and k values for each method and state which method of the three is more accurate. Discuss why you think it is more accurate.

6. Using your data from COD and ultimate BOD, calculate the non-biodegradable fraction of your samples.

7. Provide a table with your estimated BOD_5 values using the k and L_o values calculated using the 3 methods. Are these values typical of natural water samples? Note: provide a reference from the literature. Discuss any differences.

8. Determine if nitrification occurred in your samples by using your BOD_t versus *time* graphs (similar to Figure 30).

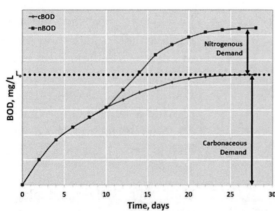

Figure 30. Carbonaceous oxygen demand and nitrogenous oxygen demand curves over time
Source: Daniel Meeroff

Liquid remaining in sample bags from the field must be disposed of by flushing down the toilet. The empty bag must be placed in the biohazardous waste container (red bag).

Viable bacteria water samples are not to be poured into a sink or placed directly in trash receptacles. Dispose of any sample containing biological seed by flushing the liquid down the toilet. All BOD bottles must be flushed with three volumes of tap water and brushed with soap and water. Then place them in the dish washer.

In case of accidental spill/exposure to samples containing seed, wash hands and lower arms thoroughly with a germicidal soap. Clean all surfaces affected with a chlorine bleach solution or a Clorox wipe.

Decontaminate the bottle containing the biological seed with 5% bleach for 30 minutes.

Use this handy chart to convert your dissolved oxygen values to T = 20°C.

T (°C)	DO$_{sat}$	0	0.1	0.2	0.3	0.4	0.5	0.6	0.7	0.8	0.9
14	10.306	0.8822	0.8841	0.8860	0.8879	0.8899	0.8918	0.8938	0.8957	0.8977	0.8996
15	10.084	0.9016	0.9035	0.9055	0.9074	0.9093	0.9113	0.9133	0.9152	0.9172	0.9192
16	9.870	0.9212	0.9231	0.9250	0.9270	0.9289	0.9308	0.9328	0.9348	0.9367	0.9387
17	9.665	0.9407	0.9426	0.9446	0.9465	0.9485	0.9504	0.9524	0.9544	0.9564	0.9584
18	9.467	0.9604	0.9623	0.9643	0.9662	0.9682	0.9702	0.9722	0.9741	0.9761	0.9781
19	9.276	0.9802	0.9821	0.9841	0.9860	0.9880	0.9900	0.9920	0.9940	0.9960	0.9980
20	9.092	1.0000	1.0020	1.0039	1.0059	1.0078	1.0098	1.0118	1.0138	1.0158	1.0178
21	8.915	1.0199	1.0218	1.0238	1.0258	1.0278	1.0298	1.0318	1.0338	1.0358	1.0379
22	8.743	1.0399	1.0419	1.0439	1.0458	1.0478	1.0498	1.0518	1.0538	1.0559	1.0579
23	8.578	1.0599	1.0619	1.0639	1.0659	1.0679	1.0699	1.0719	1.0739	1.0760	1.0780
24	8.418	1.0801	1.0821	1.0841	1.0861	1.0881	1.0901	1.0921	1.0942	1.0962	1.0983
25	8.263	1.1003	1.1023	1.1043	1.1064	1.1084	1.1104	1.1124	1.1145	1.1165	1.1186
26	8.113	1.1207	1.1227	1.1247	1.1267	1.1287	1.1308	1.1328	1.1349	1.1369	1.1390
27	7.968	1.1411	1.1431	1.1451	1.1472	1.1492	1.1513	1.1533	1.1554	1.1574	1.1595
28	7.827	1.1616	1.1636	1.1657	1.1677	1.1698	1.1718	1.1739	1.1759	1.1780	1.1801
29	7.691	1.1822	1.1842	1.1862	1.1883	1.1903	1.1924	1.1945	1.1965	1.1986	1.2007
30	7.559	1.2028	1.2049	1.2069	1.2090	1.2111	1.2132	1.2152	1.2173	1.2195	1.2216
31	7.430	1.2237	1.2257	1.2278	1.2299	1.2320	1.2341	1.2362	1.2383	1.2404	1.2425
32	7.305	1.2446	1.2467	1.2488	1.2509	1.2530	1.2551	1.2572	1.2593	1.2615	1.2636
33	7.183	1.2658	1.2678	1.2699	1.2720	1.2741	1.2762	1.2784	1.2805	1.2826	1.2848
34	7.065	1.2869	1.2890	1.2911	1.2932	1.2953	1.2975	1.2996	1.3017	1.3039	1.3060
35	6.950	1.3082	1.3103	1.3125	1.3146	1.3168	1.3189	1.3211	1.3233	1.3254	1.3276

Example of how to use the DO conversion chart

Example 5.3. Let's say that the DO probe reading is 8.31 mg/L as O_2 at 23.1°C. What is the true DO reading at 20°C?

Solution:
First, read on the chart for 23.1°C, the conversion factor is 1.0619.

Now multiply your probe reading value times the conversion factor:

$$8.31 \, mg/L \times 1.0619 = 8.82 \, mg/L$$ So, the reported DO value is 8.82 mg/L at 20°C.

Sample Data Tables

Blank w/out seed						
Day	DO	Temp.	Temp. Corr. Factor	Corrected DO	$D=DO_0-Do_t$	Flag
0	8.03	22.77	1.055504	8.48	0.00	
1	8.27	21.10	1.021119	8.44	0.03	
4	7.29	21.16	1.023188	7.46	1.02	3
5	7.13	21.42	1.027351	7.33	1.15	3
6	7.24	19.60	0.991708	7.18	1.30	3
7	6.55	20.14	1.001874	6.56	1.91	3

Seeded Blank						
Day	DO	Temp.	Temp. Corr. Factor	Corrected DO	B_0-Bt	Flag
0	7.83	23.11	1.06142	8.31	0.00	
1	7.48	20.70	1.013265	7.58	0.73	
4	5.46	20.72	1.013265	5.53	2.78	3
5	5.37	21.33	1.025265	5.51	2.81	3
6	5.44	19.80	0.995837	5.42	2.89	3
7	5.11	20.27	1.005642	5.14	3.17	3

45 mL Sample No Seed								
Day	DO	Temp.	Temp. Corr. Factor	Corrected DO	$D=DO_0-Do_t$	$P=(x/300)$	BOD_t	Reported BOD	Flag
0	9.33	23.24	1.065401	9.94	0.00	0.15	0.00	0.00	
1	8.38	21.10	1.021119	8.56	1.38	0.15	9.22	<13.3	1
4	7.93	21.06	1.021119	8.10	1.84	0.15	12.28	<13.3	1,3
5	7.60	21.33	1.025265	7.79	2.15	0.15	14.32	14.3	3
6	7.66	20.00	1.000000	7.66	2.28	0.15	15.20	15.2	3
7	7.39	20.53	1.009439	7.46	2.48	0.15	16.54	16.5	3

100 mL Sample With Seed										
Day	DO	Temp.	Temp. Corr. Factor	Corrected DO	$D=DO_0-Do_t$	$P=(x/300)$	$F=(1-P)$	B_0-Bt	BOD_t	Reported BOD	Flag
0	9.48	23.24	1.065401	10.10	0.00	0.33	0.67	0.00	0.00	0.00	
1	8.78	21.10	1.021119	8.97	1.13	0.33	0.67	0.73	1.95	<4.6	1
4	4.21	21.06	1.021119	4.30	5.80	0.33	0.67	2.78	11.94	11.9	3
5	3.37	21.33	1.025265	3.46	6.64	0.33	0.67	2.81	14.44	14.4	3
6	3.15	20.00	1.000000	3.15	6.95	0.33	0.67	2.89	15.19	15.2	3
7	2.54	20.53	1.009439	2.56	7.54	0.33	0.67	3.17	16.40	16.4	3

125 mL Sample With Seed										
Day	DO	Temp.	Temp. Corr. Factor	Corrected DO	$D=DO_0-Do_t$	$P=(x/300)$	$F=(1-P)$	B_0-Bt	BOD_t	Reported BOD	Flag
0	9.45	23.24	1.065401	10.07	0.00	0.42	0.58	0.00	0.00	0.00	
1	8.62	21.10	1.021119	8.80	1.27	0.42	0.58	1.38	1.10	<4.6	1
4	4.11	21.06	1.021119	4.20	5.87	0.42	0.58	1.84	11.51	11.5	3
5	2.68	21.33	1.025265	2.75	7.32	0.42	0.58	2.15	14.56	14.6	3
6	2.31	20.00	1.000000	2.31	7.76	0.42	0.58	2.28	15.43	15.4	3
7	1.69	20.53	1.009439	1.71	8.36	0.42	0.58	2.48	16.60	16.6	3

Note:

Flag 1. The DO deficit: ($DO_{t=0} - DO_t$) must be larger than 2.0 mg/L

Flag 2. The final DO must be larger than 1.0 mg/L

Flag 3. The blank DO deficit ($DO_{t=0} - DO_t$) must be less than 0.2 mg/L

Flag 4. The standard deviation of the samples must not exceed 30% of the mean

Sample Summary Tables

Day	125	100	28	Avg y	ln(y)	(t/y)^(1/3)	yt+1
0	0.00	0.00	0.00	0.00			
1	< 4.21	< 5.16	< 2.14				
4	< 4.30	< 5.29	< 2.14				
5	4.59	< 5.35	< 2.14	4.59	1.52	1.03	5.09
6	5.10	5.08	< 2.14	5.09	1.63	1.06	5.22
7	5.12	5.33	< 2.14	5.22	1.65	1.10	5.49
8	5.49	5.49	< 2.14	5.49	1.70	1.13	5.70
9	5.69	5.71	< 2.14	5.70	1.74	1.16	

Method	k (day^{-1})	L_o (mg/L)	BOD_5 (mg/L)	Calc. BOD_5 (mg/L)
1	0.11	34.0	14.4	14.4
2	0.18	23.2	14.4	13.9
3	0.37	17.1	14.4	14.4

Average	0.22	24.8

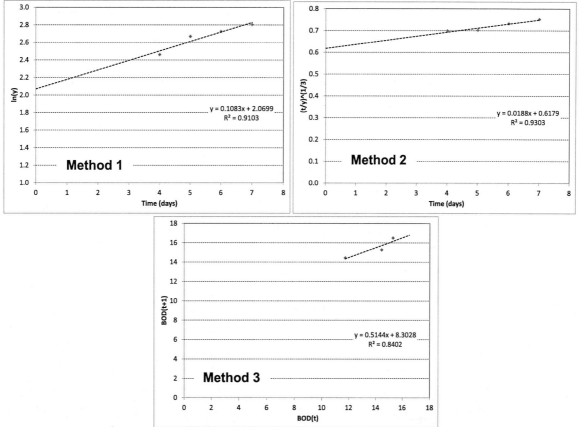

Source: Daniel Meeroff

Activity #6. DO Sag Curve

Purpose:

- To construct a Streeter-Phelps DO sag curve for an artificial wetlands environment, and understand the Streeter-Phelps model limitations and assumptions.

Background

The concentration of dissolved oxygen (DO) in a water body is an indicator of the general health of the water body. All water bodies have a mechanism of self-purification. This is known as **assimilative capacity**. If the water body exceeds its assimilative capacity, then detrimental effects will occur to the fish population in the water body, particularly if DO below 4 – 5 mg/L as O_2.

To assess the capability of a water body to absorb a waste load allocation, environmental engineers can construct a **DO sag curve** profile as shown in Figure 31.

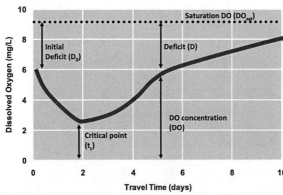

Figure 31. Example of a DO sag curve profile
Source: Daniel Meeroff

The mathematical expression for the DO sag curve takes into account the competing forces of microbial deoxygenation (from the decomposition of carbonaceous and nitrogenous oxygen demand) and reaeration from the atmosphere and photosynthesis of aquatic plants.

The model that we use is the classic Streeter-Phelps model. This model assumes the following:

1. Continuous discharge of waste (Q_W is constant)
2. Plug-flow conditions (no dispersion of wastes)
3. The water body is uniformly and completely mixed over the vertical cross section.
4. The horizontal velocity is constant over time.

The DO sag curve is given by:

$$D = \frac{k_1 L_0}{(k_2 - k_1)}(e^{-k_1 t} - e^{-k_2 t}) + D_0(e^{-k_2 t})$$

$$D = DO_{sat} - DO$$

Where,
D = dissolved oxygen deficit (mg/L)
DO_{sat} = saturated dissolved oxygen at temperature, T (mg/L)
DO = dissolved oxygen at time, t (mg/L)
k_1 = deoxygentation rate constant (day^{-1})
k_2 = reaeration rate constant (day^{-1})
L_0 = initial ultimate BOD after mixing (mg/L)
D_0 = initial dissolved oxygen deficit after mixing (mg/L)
t = time of travel (days)

The critical point (t_c) is given by the following equation:

$$t_c = \frac{1}{k_2 - k_1} \ln\left[\frac{k_2}{k_1}\left(1 - D_0 \frac{k_2-k_1}{k_1 L_0}\right)\right]$$

Pre-Lab Questions

1. The ultimate BOD of a river after mixing with a sewage outfall is 75 mg/L, the DO after mixing is 7.0 mg/L at 27°C, k_1 is 0.30/day and k_2 is 0.95/day. The river is flowing at 48 miles per day. The only source of BOD in the river is this single outfall.

    a) Plot DO as a function of distance downstream from the outfall in 5 mile increments.

    b) Find t_c (in days) and x_c (in miles).

    c) Report the minimum DO (in mg/L).

    d) What fraction of BOD would have to be removed from the sewage to assure a minimum DO of 5.0 mg/L everywhere downstream?

Procedure

1. **Collect samples from your assigned sampling point(s).** Team 1 will collect 1-L of sample from the mixing point of the outfall and the surface water body to be used as the value of L_0. Team 1 will also measure the dissolved oxygen concentration and temperature at this point. This will be the DO_0 value. **NOTE: If your group is not Team 1, then you must obtain the L_0 and DO_0 data points from Team 1.**

    • Each team will be assigned additional sampling points downstream and will collect a 1-L sample.

    • The 1-L sample is collected using a sampling pole. Rinse the cup holder with three volumes of water from your

site and dispose of the rinsate downstream. On the fourth volume, pour the sample through a coffee filter (that you bring with you to the field) into a sterile sample container. Make sure the container is sealed securely and then store it upright in the cooler with ice.

• Label the sample container with the sample number, team name, and time collected using a Sharpie.

• At each sampling location, collect the following data:

    o Average depth (h) of the water body in meters
    o GPS coordinates
    o Dissolved oxygen concentration (mg/L as O_2)
    o Temperature (°C)

The DO probe is calibrated in the laboratory prior to field sampling. Do not attempt to calibrate in the field.

2. **Determine the stream velocity.** Estimate the average horizontal velocity (u) in the water body (in m/s) by dividing the incoming flowrate (Q) by the cross-sectional area of the water body (*depth × width*) at your sampling point determined by estimation from an aerial photograph. If Q is not known, then measure the stream velocity by measuring how long it takes for an object to traverse a known distance.

3. **Provide a table like the one below.** Share your table with all of your data points to your other classmates so that they can plot the Streeter-Phelps model.

Reading #	Distance from Outfall (m)	Depth (m)	DO (mg/L)	Temp (°C)	Velocity (m/s)
1	37.5	0.5	3.30	23.8	0.055
2	53.0	0.7	3.84	24.1	0.042
3	98.5	0.9	3.99	24.0	0.014

Discussion Questions

1. How can you improve the estimate of the average stream velocity?

2. Estimate $k_2 = \dfrac{3.9u^{1/2}}{h^{3/2}}$ for all samples. How reasonable were your values?

3. Explain the shape of your DO sag curve. Why do you think it turned out this way?

4. Once the BOD experiment is complete and you have all of the data points from all of the groups, plot your points and estimate the value of the critical point (t_c). On your graph, show the DO_{sat} and the minimum DO level. Take your values for D_0, L_0, and t_c and k_2. Now back calculate the value of k_1 using the t_c equation. Is this value reasonable?

Proper Safety and Disposal

Liquid remaining in sample bags from the field must be disposed of by flushing down the toilet. The empty bag must be placed in the biohazardous waste container (red bag).

Viable bacteria water samples are not to be poured into a sink or placed directly in trash receptacles. Dispose of any sample containing biological seed by flushing the liquid down the toilet. All BOD bottles must be flushed with three volumes of tap water and brushed with soap and water. Then place them in the dish washer.

In case of accidental spill/exposure to samples containing seed, wash hands and lower arms thoroughly with a germicidal soap. Clean all surfaces affected with a chlorine bleach solution or a Clorox wipe.

Decontaminate the bottle containing the biological seed with 5% bleach for 30 minutes.

Activity #7. Solid Waste Facility Field Trip and Waste Audit

Purpose:

- To understand how different engineered processes are utilized in an integrated solid waste management operation.

- To determine factors that impact the composition and quantities of waste generated.

Background

A **waste audit** is a formal, structured process used to quantify the amount and types of waste being generated. Information from audits helps identify current waste practices and how they can be improved. The ultimate goal of a waste audit is to identify what materials are being discarded to come up with solutions for better waste management.

In order to accomplish this goal, a waste audit will be conducted using an example waste audit data sheet (Table 7). The column labeled "Tare" is the empty weight of the sub-container for the category. The column labeled "Gross" is the weight of the sorted waste plus the sub-container. The column labeled "Weight" is the difference between Gross and Tare. Some typical waste audit sub-categories include but are not limited to the following:

1. **Paper**

   - **Cardboard**: packing boxes, cereal boxes, gift boxes, beer/soda paperboard boxes, corrugated cardboard, polycoated drink boxes, juice cartons, milk cartons.

   - **Other recyclable paper products**: newspapers, magazines, catalogs, telephone books, printer paper, copier paper, mail, all other office paper without wax liners.

   - **Non-recyclable paper products**: all other paper with wax coatings and sanitary paper.

2. **Plastic containers**

   - **Recyclable plastic:** Containers with the triangle label symbol 1 – 7.

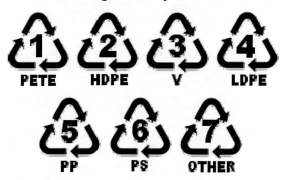

© shutterstock.com

   - **Non-recyclable plastic**: plastic bags, packaging, caps, lids, and any plastic with numbers above #7 or without a number.

3. **Glass** (Ex. clear, brown and green glass food and beverage containers).

4. **Metals**

   - **Non-ferrous**: aluminum and tin

   - **Ferrous**: steel food and beverage containers

5. **Organics**

   - **Kitchen/food scrapings**

   - **Leafy yard waste**

6. **Hazardous** (Ex. garden sprays, medicines, batteries, mineral oil, paint, aerosols, light bulbs, etc.)

7. **Miscellaneous** (Ex. electronics, rubber, leather, textiles, etc.)

Table 7. Sample waste audit checklist.

Waste Audit Checklist

Date: _____
Time: _____
Location: _____
Performed by: _____

CATEGORY	DESCRIPTION	TARE	GROSS	WEIGHT (lbs)
Paper	Cardboard			
	Other Recylable Paper			
	All Other Non-Recyclable			
	TOTAL PAPER			
Plastic	Type 1-7 Recyclable			
	All Other Non-Recyclable			
	TOTAL PLASTIC			
Glass	Green, Clear, Brown Recyclable			
	All Other Non-Recyclable			
	TOTAL GLASS			
Metal	Ferrous			
	Non-Ferrous			
	TOTAL METAL			
Organic	Kitchen/Food			
	Leafy Yard Waste			
	TOTAL ORGANIC			
Hazardous	All			
	TOTAL HAZARDOUS			
Misc.	Other			
	TOTAL MISC			
	TOTAL WASTE			

NOTES _____
_____
_____
_____
_____
_____
_____
_____
_____
_____
_____
_____
_____
_____
_____
_____
_____

Source: Daniel Meeroff

There are many factors that impact how much and what types of wastes are generated at a particular site. Some of these factors include: climate, location, affluence, number of people in the household, square footage of the house, number of pick-ups per week, household income, pick-up days, behavioral issues, age, number of hours per week at home, distance from landfill, etc.

Pre-Lab Questions

1. Make a table listing each team member's name, where the person lives, where the person's garbage goes after it is collected, where the person's recyclable material goes after it is collected, and estimated recycling percentage of each team member's garbage.

2. Create your own waste audit data sheet categories for your experiment and submit digital photographs of each of the different categories of waste items to be sorted in the waste audit.

3. How much waste do you expect to collect per person per day? Don't forget to provide a source for the value above. Using this value, what is the total amount of material that you expect to collect in one week?

Procedure

1. **Plan the audit carefully.** Good planning is essential to ensuring the audit goes smoothly. Make sure that your plan does not affect the household behavior, but captures all waste.

2. **Collect the waste.** Each team member should collect their own solid waste for a period of 7 consecutive days.

© Shutterstock.com

- Weigh an empty sorting bin or bag, and record the weight on the outside of the bin or bag and on your data sheet. Each day, sort the waste into the appropriate bin (e.g., glass, plastics, metal, etc.). Then, at the end of each day, weigh each one and add up the weights for each category for the week. Do not store the material longer than necessary.

- Record your weights on your data sheet before disposing of the material (or recycling it, if appropriate).

- Take photos and make note of any relevant observations, as you sort.

Source: Daniel Meeroff

- Organic waste should not be kept for more than three days.

- **Analyze the data and report the results.** Once all of the waste has been sorted by categories, and the weights have been recorded in your data sheet, you should be able to generate a pie chart that shows the percentages of each category similar to the one shown in Figure 32.

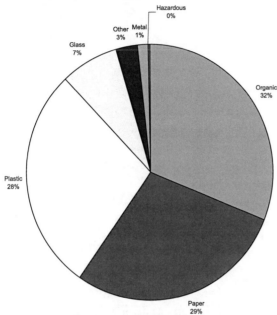

Figure 32. Pie chart of waste categories for the Standard Hotel, Miami Beach, FL
Source: Daniel Meeroff

Discussion Questions

1. Provide the location of the facility (physical address), map (or aerial photograph), and size of the facility.

2. Provide a brief synopsis of the essential background to understand how the facility works. (i.e. how does it work, what does it do, why was it built, what are its benefits, what are its disadvantages, and what are the engineering alternatives).

3. What is the capacity of the facility and the expected lifetime? What would you change in a typical sanitary landfill to make it last longer?

4. Why is daily cover necessary on a sanitary landfill? Where does SWA get their daily cover fill from? What is its composition?

5. What types of materials are collected at the recycling facility? Make a table showing which materials make the most money and which ones lose the most money. Discuss.

6. What is leachate? What happens to the leachate generated at this facility? Are there any uses for this material?

7. Using the actual waste generation rate from this facility, if we built a landfill in the shape of a trapezoidal prism on 88 acres, with one lift of 24 ft high, side slope of 30 degrees, compacted specific weight of solid wastes in the landfill of 750 lb/yd^3, how many years will it last?

8. Plot a pie chart showing what percentage of each of your team member's waste, by weight, is recyclable. Discuss.

9. If your sample constitutes the average waste generated in one week, how much recyclable waste could be diverted from the landfill **each year**, if 70% of the recyclable material actually got recycled?

10. Calculate the amount of waste generated per person per capita day. How does this compare to the average amount per person per capita day in the U.S.?

11. What do you recommend to reduce the amount of waste from your own home that is going to the landfill?

Proper Safety and Disposal

Dispose of all of your sorted material in the waste container or recycling bin, as appropriate.

Activity #8. Global Environmental Issues

Purpose:

- To observe the effect of surface color on local temperatures using a simple greenhouse model.

Background

One of the main reasons that life is able to exist on our planet is because the temperatures over most of the Earth's surface are within the tolerable limits for living things. The energy that warms our planet comes from the sun in the form of electromagnetic radiation, specifically visible and ultraviolet light. As light passes through the Earth's atmosphere, several mechanisms govern the amount of energy that is received at the surface.

- Some of the energy is absorbed by the atmosphere or reflected back into space and never reaches the surface
- The remaining portion, which is approximately half of the incoming radiation, reaches the surface

Some of the sun's transmitted energy gets absorbed by features on the Earth's surface (e.g. trees, mountains), while the remaining portion is re-radiated back by the Earth. Much of this re-radiated energy is in the form of heat (infrared radiation). If this heat energy were lost, the temperatures would be extremely cold (e.g. average global temperatures of -18°C), but certain greenhouse gases in the Earth's atmosphere trap this heat energy and reflect it back to the surface to maintain the planet's warmth, just like the way a greenhouse keeps plants warm during the winter. This is called the "**Greenhouse Effect.**"

The Greenhouse Effect is a physical process that occurs due to the presence of gases in our atmosphere, such as water vapor, carbon dioxide, methane, and nitrous oxide, which act like the glass in a greenhouse, trapping the escaping heat and re-radiating it back towards the surface in the form of infrared energy, or heat.

The Greenhouse Effect also evens out the temperatures around the world by distributing the heat energy, so that the side of the Earth that is away from the sun (night side) does not get extremely cold, and the side facing the sun (day side) does not get extremely hot. This delicate balance depends on the composition of the greenhouse gases in the atmosphere. However, human activities are pumping more and more carbon dioxide into the air (Figure 33) due to our increasing energy, housing, and agricultural demands. For example, fossil fuel combustion produces an additional burden of greenhouse gases, animal farming produces large amounts of methane, and CFCs (e.g. chlorofluorocarbon refrigerants such as Freon), all contribute to additional pressures on the Earth's natural Greenhouse Effect.

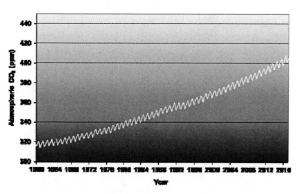

Figure 33. Keeling curve of atmospheric CO_2 from the Mauna Loa Observatory from 1958 – 2017 (Adapted from Keeling et al. 2001)

Scientists have monitored measurable increases in global temperatures over a relatively short period of time (Figure 34), in which temperatures were decreasing until the late 1800s, after which the average temperature climbs rapidly. If the increasing temperatures continue to rise at the current rate, this could have devastating effects to the global climate.

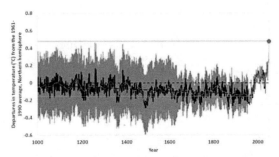

Figure 34. Long-term northern hemisphere departures from the 1961-1990 average temperature using tree rings, corals, ice cores, and historical trends 50-year averages. Most recent data is year-by-year thermometer-based (Adapted from Mann et al. 1998)

For instance, this warming trend may cause glaciers and polar ice caps to melt, leading to rising sea levels that would flood out coastal communities. In addition, weather patterns would be altered causing natural disasters which could result in many animal or plant species being driven to extinction.

Procedure

1. **Place the greenhouse model in a sunny location,** where all parts of the model receive the same amount of sunlight (no shaded areas). Then place a clear plastic cover over the top and insert a clear base filter. Turn the model so that the cover faces the sun, and make sure that thermometers are not shaded by any portion of the model.

2. **Record baseline temperatures.** Place two temperature recording devices so that the

tips are resting against the white base area of the model. Allow 2-3 minutes to equilibrate, then observe both readings until they are both consistent. This is the baseline temperature.

Courtesy of Florida Atlantic University

3. **Record inside temperature.** Insert the other thermometer so that it is completely inside the greenhouse area.

4. **Record outside temperature.** Insert one thermometer so that it is sticking outside the greenhouse area.

5. **Begin taking temperature readings in two-minute increments.** Continue for a total of 8 minutes.

6. **Repeat temperature readings with a red base filter.**

7. **Repeat temperature readings with an orange base filter.**

8. **Repeat temperature readings with a blue base filter.**

9. **Record your data in a table.** Label as Effect of local geography on temperature, similar to the one that follows:

Temperature (°C)							
	Clear		Red		Orange		Blue	
Min.	In	Out	In	Out	In	Out	In	Out
2								
4								
6								
8								

10. Repeat temperature readings with clear cover filter.

11. Repeat temperature readings with red cover filter.

12. Repeat temperature readings with orange cover filter.

13. Repeat temperature readings with blue cover filter.

14. Record your data in a table. Label as Effect of atmosphere on temperature.

Discussion Questions

1. What information does this experiment give you about the effect of a local geographic area on temperature in that area?
2. What are the different colored cover filters models of?
3. Which filter most accurately represents our atmosphere and why?
4. What combinations of filters would you use in the Arctic? Open ocean? Rainforest?
5. What conditions do you predict will cause the greatest temperature increase? What about the least temperature increase?

Activity #9. Mass Balance/Dilution

Purpose:

- To test the mass balance concept of a conservative pollutant in a membrane filtration setup.
- To determine if Beer's Law applies to a particular solution using spectroscopic methods

Background

Spectroscopic methods of analysis can be used to determine the concentration of many dissolved substances. The Beer-Lambert Law, also known as Beer's Law, states that for a given solution, absorbance of light energy is directly proportional to its path length and the concentration of the absorbing substance:

$$\log \frac{I_o}{I} = A = abC$$

Where:

I = intensity of monochromatic light transmitted through the test solution

I_o = intensity of light transmitted through the reference solution (i.e. blank)

A = absorbance (dimensionless)

a = absorptivity, a constant for a given solution/system and a given wavelength

b = light path length (cm)

C = concentration of solute, g/L

Spectroscopy involves the instrumental measurement of light intensity. In this analysis, the light transmitted through the solution is measured. The transmittance of a solution (T) is defined as I/I_o, and $\%T$ as $I/I_o \times 100$. The absorbance is $1/T$.

The spectrophotometer is an analytical instrument that makes possible a quantitative measurement of light passing through a clear solution. This instrument is capable of supplying light with a narrow wavelength bandwidth and is equipped with sensitive detectors to measure light intensity. The instrument's light source is dispersed by a diffraction grating (or prism) and the desired wavelength region is selected by passing through a slit and a monochromator to generate a narrow wavelength bandwidth of about 20 nm. The selected wavelength band is then passed through the sample. Some of the light is absorbed by constituents in the sample. All of the light not absorbed by the solution is received by the detector, the signal from which is displayed on the instrument scale as transmittance or absorbance.

The sensitivity of analysis for a particular solution and the degree of adherence to Beer's Law depends on the choice of wavelength. Wavelength selection is based primarily on an evaluation of the solution absorption spectrum that describes the absorption characteristics as a function of wavelength (Figure 35).

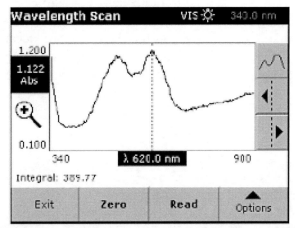

Figure 35. Screenshot of a full spectrum scan
Source: Daniel Meeroff

Maximum sensitivity, or the largest absorptivity (a), is found at wavelengths where maximum light absorption occurs. Other portions of the spectrum may be used if a reproducible curve of absorbance versus concentration can be obtained.

Deviations from Beer's Law typically occur at higher concentration ranges. That is why it is so important to establish a calibration curve and check the correlation coefficient value to make sure that it is in the linear range.

For example, five standards of known concentration are placed in the spectrophotometer and their absorbance values are recorded. Next, these values are plotted (Figure 36), and the equation of the line ($y = 0.032x$) and the correlation coefficient ($r^2 = 0.991$) are noted.

Now that we have established our standard curve, if we measure an unknown sample with absorbance of 0.30 units, then we can calculate the concentration in the sample by setting $y = 0.30$ and solving for x, as follows:

$$y = 0.032x$$
$$0.30 / 0.032 = 9.3 mM$$

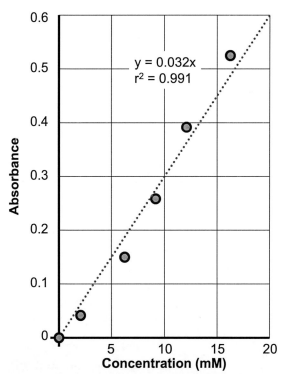

Figure 36. Sample calibration curve showing absorbance v. concentration
Source: Daniel Meeroff

Membrane filtration is a physical process that requires an applied pressure to force clean water through a porous membrane, against the natural direction of flow. The clean water stream is called, the ***permeate***, and the process creates a brine stream called the ***concentrate***.

Depending upon the pore sizes and the applied pressures, membrane systems can be designed to target specific contaminants (Table 8).

Table 8. Summary of membrane filtration characteristics

Membrane Designation	Pore size µm	Pressure psi	Major Function
Reverse Osmosis (R/O)	0.0001 – 0.001	50 – 1000	Removes monovalent ions 99.4% rejection possible
Nanofiltration (NF)	0.001 – 0.01	20 – 400	Removes TOC and DBPs Removes divalent cations
Ultrafiltration (UF)	0.01 – 0.1	5 – 55	Removes colloids (viruses too) Removes a fractional amount of TOC
Microfiltration (MF)	0.1 – 2.0	5 - 35	Removes TSS Removes bacteria, protozoan cysts

The design of membrane systems can be determined using a simple one input-two output mass balance model. For this experiment, we will use a bench scale ultrafiltration unit (Figure 37).

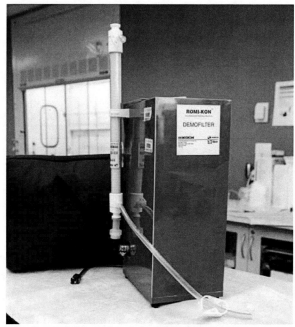

Figure 37. Bench scale ultrafiltration unit
Courtesy of Florida Atlantic University

Let's do an example to make sure that you understand how to compute the values.

> *Example. A membrane system produces 19,000 m^3/d of treated water. The membrane recovery is 75%, such that $Q_o = 19,000 \div 0.75 = 25,333\ m^3/d$. The initial concentration of salt is 1000 mg/L and the unit produces a permeate with 67 mg/L of salt. What is the concentration of salt in the concentrate stream?*

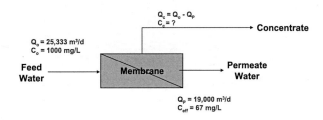

First, determine the flow rate in the concentrate stream:

$Q_c = Q_o - Q_P = 25,333 - 19,000 = \underline{\textbf{6,333 m}^3\textbf{/d}}$

Then, determine the concentration C_c using mass balance:

$$Q_cC_c = (6,333)C_c = Q_oC_o - Q_PC_{eff}$$

$$(6,333)C_c = (25,333)(1000) - (19,000)(67)$$

$$\underline{C_c = \textbf{3800 mg/L}}$$

Pre-Lab Questions

1. Can a soluble substance be analyzed spectrophotometrically if its light absorption varies with concentration, but it does not obey Beer's Law?

2. If a solution has a red color, what wavelength (in nanometers) and color of light is being absorbed by the solution?

3. What does it mean if the standard curve has a correlation coefficient less than 0.50?

1. **Obtain an appropriate sample.** Examples of possible substances include transmission fluid, milk, or some other fluid designated by the instructor in water.

2. **Make 5 dilutions.** Use a high concentration like 2.5 – 10% to start.

3. **Determine the maximum absorption peak.** Use the spectrophotometer on full scan mode, and record the wavelength with the maximum absorbance value. Take a photograph of the full scan.

4. **Create a standard curve.** Using 5 appropriate dilutions (i.e. 0.00%, 0.25%, 0.50%, 1.00%, 1.67%, 2.50%), measure the absorbance values on the spectrophotometer, and plot a graph of absorbance v. concentration, showing the equation of the line and the correlation coefficient, as follows:

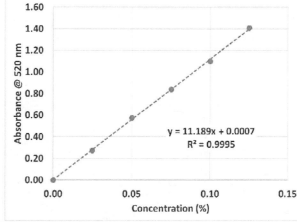

Source: Daniel Meeroff

5. **Use the membrane filtration unit.** On steady state mode to filter the sample for 15-60 minutes.

6. **Measure the flowrate of the permeate stream and the concentrate stream.** Use a graduate cylinder and a stopwatch. Repeat these measurement 3 – 5 times

each and record the individual values, the average, and the standard deviation in a table.

7. **Determine concentrations.** Collect subsamples of the permeate, the feed, and the concentrate streams. Then analyze them spectrophotometrically. Calculate the concentrations using the equation from your standard curve in step 4. Dilute as necessary to remain in the quantitation range.

8. **Calculate the feed concentration using mass balance.** With the flowrate and concentration values recorded for the permeate and the concentrate streams, calculate the concentration of the feed with the mass balance equation.

- Make a table with the known concentration of feed, the measured concentration of feed, and the calculated concentration of feed.

- Determine the percent errors of your feed concentration obtained by measurement and by calculation compared to the actual value.

Discussion Questions

1. Show a photograph of the full spectrum scan of your sample. Assuming that more than one peak or plateau occur in an absorption spectrum, what would be the advantages, if any, in choosing a wavelength at an absorbance peak other than the maximum?

2. Show a graph of the standard curve with the equation of the line and the r^2 value. Did your standard curve follow Beer's Law? Explain.

3. Make a table showing the value of the feed as estimated by mass balance, as measured by spectrophotometry, and as computed from the original sample dilution. Then list the percent errors. Explain what might have caused any observed differences in the feed concentration measurement.

4. Explain why you think the permeate is safe or unsafe to drink after treatment.

Proper Safety and Disposal

Dispose of all of your sample and all standards in the appropriate waste container.

Activity #10. Solids Analysis and Conductivity

Purpose:
- To measure the TS, TVS, TSS, VSS, and TDS content of environmental samples using the gravimetric method and compare to the data obtained using a conductivity estimation using a probe.

Background

Solids analysis is useful in characterizing water quality of natural and treated waters. Too much solids can cause undesirable cloudiness and excessive sedimentation. They can also harbor microbial pathogens and clog filters. Solids are classified as shown in Figure 38.

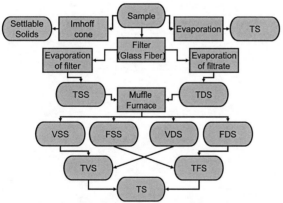

Figure 38. Breakdown of solids content in a water sample
Source: Daniel Meeroff

Total Solids (TS): total amount of solids found after drying a sample at 105°C and is determined as follows:

$$TS = \frac{Dry\ wt. - Container\ wt.}{Sample\ Volume}$$

Total Volatile Solids (TVS): solids that incinerate at 550°C, which is an indicator of organic content and is determined as follows:

$$TVS = \frac{Dry\ wt. - Burned\ wt.}{Sample\ Volume}$$

Total Suspended Solids (TSS): solids caught on a 0.45 µm filter and is determined as follows:

$$TSS = \frac{Dry\ filter\ wt.(+pan) - Clean\ filter\ wt.(+pan)}{Sample\ Volume\ (filtered)}$$

Volatile Suspended Solids (VSS): solids caught on a 0.45 µm filter that incinerate at 550°C, which is an indicator of suspended organic content and is determined as follows:

$$VSS = \frac{Dry\ filter\ wt.(+pan) - Burned\ filter\ wt.(+pan)}{Sample\ Volume\ (filtered)}$$

Total Dissolved Solids (TDS): solids dried at 105°C that pass the 0.45 µm filter and is determined by as follows:

$$TDS = \frac{Dry\ wt.(+pan) - pan\ wt.}{Sample\ Volume\ (filtered)}$$

Note that TS should be equal to the TSS plus the TDS.

Let's do an example to make sure that you understand how to compute the values.

Example. A 100 mL sample weighs 1.6346 g after drying. The dish weighs 1.6084 g. What is the concentration of total suspended solids in the sample?

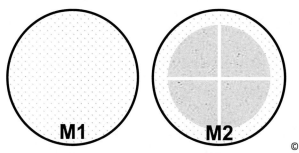

Source: Daniel Meeroff

Given:
- (M1) Mass of filter after drying @105°C = 1.6084 g
- (M2) Mass of filter plus residue after drying @105°C = 1.6346 g

Find:
TSS = ?

Expected:
0.250 g/L

Solution:
TSS = (M2 – M1)/Volume
TSS = (1.6346 – 1.6084)/0.100 = <u>0.262 g/L</u>

Conductivity is a measure of the ability of solution to conduct an electric current and is reported as Siemens/cm (S/cm). Conductivity of a solution is found by measuring the conductance with a conductivity meter and conductivity cell. The conductivity (k) and conductance (λ) are related by the electrode cell constant (C), which is influenced directly by the separation distance and inversely by the area of the electrodes. If the cell constant is known, the conductivity can be determined:

$$Conductivity\ (k) = C\lambda$$

C = cell constant (cm^{-1})
λ = measured conductance (mS)

Typical values for conductivity and cell constants are found in Table 9. Conductivity is a rough measure of the relative concentration of dissolved salts. For instance, there is an approximate conversion available for conductivity and total dissolved solids (TDS):

$$TDS\,(mg/L) = c\left(\mu S \cdot L/cm \cdot mg\right) \times k\left(\mu S/cm\right)$$

Where c varies from 0.50 – 0.65, typically.

Table 9. Conductivity of typical aqueous solutions and recommended cell constants at 25°C

Sample Type	Conductivity mS/cm	Cell Constant cm^{-1}
Pure water	0.05	0.1
Distilled water	0.5	0.1
Deionized water	0.1 - 10	0.1
Demineralized water	1 - 80	0.1
Drinking water	0.5 - 1.0	1.0
Domestic wastewater	0.9 - 9	1.0
Brackish water	1 - 80	1.0
Ocean water	53	10

Pre-Lab Questions

1. Look up typical values for TSS and conductivity for natural water samples.

2. If TSS = 18.0 mg/L and TDS = 225 mg/L, what do you expect the total solids content to be in grams per liter?

3. Why do we use filter disks made of glass microfibers instead of paper filters?

4. What does it mean if the TSS value is less than the VSS value?

Procedure

1. **Obtain sample.** Each group should provide **two** environmental samples. You need to bring **2 liters** of each sample. One sample should be rain water. If you cannot collect a rain water sample, try one of these: river water or canal water. The second sample should be seawater. A third sample (QA/QC) will be provided. Note the location and time collected for each sample.

2. **Measure Total Solids (TS).** Weigh an empty aluminum dish or ceramic crucible to the nearest 0.01 g. and record this value.

- Mix sample thoroughly, then measure 15.0 mL using a graduated cylinder and transfer into your pre-weighed dish.
- Place the dish in a drying oven at 105°C overnight until dry.
- Remove from oven and cool the dish in the desiccator for one hour to bring to room temperature. Avoid handling with your hands.
- Weigh the dish with residue to the nearest 0.01 g and record this value to determine TS.

3. **Measure Total Volatile Solids (TVS).** Place dish in a muffle furnace at 550°C to ignite the residue for 15 minutes.

- Remove from furnace and cool the dish in the desiccator for one hour to bring to room temperature. Avoid handling with your hands.
- Weigh the dish with filter residue to the nearest 0.01 g and record this value to determine TVS.

4. **Measure Total Suspended Solids (TSS).** Weigh glass microfiber filter with aluminum pan together to the nearest 0.01 g and record this value.

- Prepare filter by wetting with a small amount of reagent grade water. Turn on the vacuum pump. Then discard any water that collects in the filtering flask.
- Mix sample thoroughly, dispense 75 mL to a graduated cylinder, then pour contents onto the filter with the vacuum on, collecting the filtrate in the filtering flask.
- Disengage vacuum.
- Collect the filtrate in a graduate cylinder for the TDS procedure.

- With empty filter flask, reapply vacuum and rinse filter with 10 mL reagent water, three times.
- Disengage vacuum and remove filter with forceps.

Courtesy of Florida Atlantic University

- Place the filter in its companion aluminum dish and place in a drying oven at 105°C for 1 hour.
- Remove from oven and cool the dish in the desiccator for one hour to bring to room temperature. Avoid handling with your hands.
- Weigh the dish with filter residue to the nearest 0.01 g and record this value to determine TSS.

5. **Measure Volatile Suspended Solids (VSS).** Place the dish with filter residue from step 4 in a muffle furnace at 550°C to ignite the residue for 15 minutes.

- Remove from furnace and cool the dish in the desiccator for one hour to bring to room temperature. Avoid handling with your hands.
- Weigh the dish with filter residue to the nearest 0.01 g and record this value to determine VSS.

6. **Total Dissolved Solids (TDS, not filterable).** Weigh an empty aluminum

dish or ceramic crucible to the nearest 0.01 g. and record this value.

- Mix filtrate from step 4 thoroughly, then measure 50.0 mL using a graduated cylinder and transfer into your pre-weighed dish.
- Using a calibrated conductivity probe, record the conductivity, specific conductance, temperature, and estimated TDS of the filtrate.
- Place the dish with filtrate in a drying oven at 105°C overnight until dry.
- Remove from oven and cool the dish in the desiccator for one hour to bring to room temperature. Avoid handling with your hands.
- Weigh the dish with residue to the nearest 0.01 g and record this value to determine TDS.

7. **Calculate TS, TVS, TSS, VSS, and TDS (show calculations).** Provide a table similar to the one that follows:

	Rain Water	Sea Water	Dilution Water or QA/QC Sample
TS			
Mass of dish (M1) in grams			
Mass after 105°C (M2) in grams			
Volume (V1) in liters			
TS = (M2-M1)/V1			
TVS			
Mass after 550°C (M3) in grams			
TVS = (M2-M3)/V1			
TSS			
Mass of filter+dish (M4) in grams			
Mass after 105°C (M5) in grams			
Volume (V2) in liters			
TSS = (M5-M4)/V2			
VSS			
Mass after 550°C (M6) in grams			
VSS = (M5-M6)/V2			
TDS			
Mass of dish (M7) in grams			
Mass after 105°C (M8) in grams			
Volume (V3) in liters			
TDS = (M8-M7)/V3			

Discussion Questions

1. What is the difference between conductivity, specific conductance, and TDS?

2. Determine if TS = TDS + TSS in all of your samples by listing the TS, TSS, calculated TDS from the probe, measured TDS (gravimetrically), and relative error, and then explain the cause of any discrepancies between the calculated and the measured TDS values.

3. Make a table with the TDS probe readings and the gravimetric TDS values for all of your samples (also include the percent errors). Explain any differences in your results and determine which method is preferable (justify your answer!).

4. Did your data conform to the Langelier (1936) approximation which states that:

$$TDS\,(mg/L) = 0.64 \times specific\ conductance\,(\mu S/cm)$$

Plot your data on a graph of TDS vs. specific conductance to compute the value of c, and discuss the differences between the relationship that you found with your data and the Langelier approximation (c = 0.64).

5. Sometimes in wastewater treatment applications, technicians use the VSS value to approximate the number of live microorganisms in the aeration basin. Discuss the advantages and disadvantages of this approach.

6. Why is it important to dry the filter disk at 105°C? Why is it important to let the filter disk come to room temperature in a desiccator before making your final weight measurement?

7. What would happen if we used a filter disk with a nominal pore size of 1.2 μm instead of 0.45 μm?

Proper Safety and Disposal

Dispose of all of your samples and all standards in the appropriate waste container.

Activity #11. Turbidity

Purpose:
- To determine the amount of light scattering in environmental samples.

Background

Turbidity is an indicator of the cloudiness of a water sample measured by the amount of light scattered by the particles in suspension, such as clay, slit, colloids, and microorganisms (see Figure 39).

Figure 39. Example of a turbid sample (right) and a clear sample (left)
Courtesy of Florida Atlantic University

Modern turbidity instruments measure the light scattered at an angle to the incoming light beam and then relate this off-axis scattered light to the sample's actual turbidity. Most instruments of this type measure the orthogonal scatter ($90° \pm 30°$) because this angle is considered to be the least sensitive to particle size. These types of instruments are referred to as Nephelometers. The measurement unit is *Nephelometric Turbidity Units* (NTU).

Turbidity is used as a rough indicator of bacterial contamination. Treated drinking water must have a turbidity value lower than 0.5 NTU.

Pre-Lab Questions

1. What does turbidity mean?
2. Look up typical values for turbidity for natural water samples in your area (provide references).
3. If one sample has a TSS = 18.0 mg/L and another sample has a TSS = 22.5 mg/L, which do you expect to have the higher turbidity and why?

Procedure

1. **Calibrate nephelometer.** Using the two lowest turbidity standards, follow the manufacturer's instructions for instrument calibration. Be sure to record the instrument reading of your standards.
2. **Fill clean sample cells with samples.** Dry the outside with a Kim-wipe, handle the cell only by the cap (to avoid fingerprints), and insert the cell into the instrument with the proper alignment. Make sure the cell cover is closed after the sample cell is inserted. Allow bubbles to rise out of the light path (few seconds).
3. **Record the turbidity value in a table.** Just like the one that follows:

Sample	Turbidity (NTU)
Dilution Water	
Rain Water	
1:1 Rain Water	
Sea Water	
1:1 Sea Water	

Discussion Questions

1. Provide a table of all of your samples and the turbidity readings. Discuss differences in turbidity between your samples. What is the cause? What does it mean?

2. When you diluted one of your samples 1:1 with sterile filtered dilution water, was the turbidity value reduced by exactly half?

3. Plot your turbidity results vs. TSS and specific conductance from the previous activity. Is the mathematical relationship between turbidity and TSS and/or specific conductance a constant value for your three samples? If you wanted to relate your turbidity results to the results of your solids analysis, which parameter(s) would you use, and why?

Proper Safety and Disposal

Dispose of all of your samples and all standards in the appropriate waste container.

Activity #12. Hardness

Purpose:
1. To measure the calcium and magnesium hardness concentration in water to determine if hardness correlates with the height of soap bubbles.

Background

Water hardness is a property caused by multivalent cations, primarily calcium (Ca^{2+}) and magnesium (Mg^{2+}) ions. The total hardness is defined as the sum of the multivalent cations in a water sample, expressed in units of milligrams per liter as calcium carbonate (mg/L as $CaCO_3$).

When we say water is hard or soft, it does not mean that the water feels hard or soft when you touch it. We call water "hard" when it has high concentrations of minerals (like calcium and magnesium) dissolved in it. Table 10 lists the various classes of hardness.

Table 10. Classes of hardness (Tchobanoglous and Schroeder 1985)

Hardness Range (mg/L as $CaCO_3$)	Description
0-50	Soft
51-150	Moderately Hard
151-300	Hard
>301	Very Hard

In domestic use, hardness affects the amount of soap needed to produce a lather. Hard waters can also create *scale* in heaters and hot water pipes or soap scum in the sink, shower or bathtub, which is caused by calcium carbonate and magnesium hydroxide precipitating out at elevated temperatures. Hard water is also more difficult to clean with because the soap gets attached to the minerals in the water instead of the dirt.

Two methods are available for measurement of hardness. The first method is calculation of hardness by performing a mineral analysis on the sample and computing the hardness as the sum of the calcium and magnesium concentrations. This method is highly accurate. The second method involves forming chelates with metal ions in a titration. The most commonly used chelating agent is ethylene diamine tetraacetic acid (EDTA). Titration with EDTA provides a rapid analysis of water hardness. The chemical structure of EDTA is shown in Figure 40.

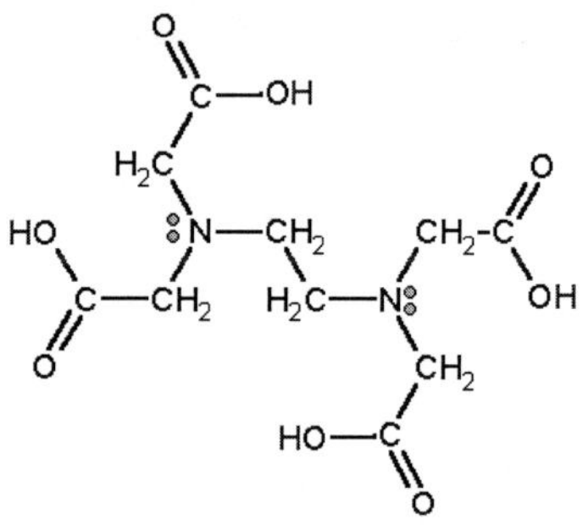

Figure 40. Chemical structure of EDTA
Source: Daniel Meeroff

When determining hardness levels using EDTA titrations, the sample is buffered at pH 10.0 to prevent precipitation of the metal ions. Since complexes of calcium-EDTA or magnesium-EDTA are not colored, a second chelating agent, Eriochrome black T (EBT) is used for the endpoint determination. A small amount of EBT is added to the test solution before the titration with EDTA. The dye forms a weak bond with the Ca^{2+} and Mg^{2+} ions, which turns the solution a red color. As the EDTA titrant is added to the solution, all of the free Ca^{2+} and Mg^{2+} ions are complexed by the EDTA, which disrupts the complex formed with the EBT dye. This is because EDTA can form a more stable complex than the dye can. This action frees the dye, and the red color changes to blue, signaling the endpoint of the titration.

Let's do an example to make sure that you understand how to compute the hardness titration.

Example. A 50.00 mL sample takes 10.68 mL of 0.0800M EDTA to reach the endpoint What is the molar concentration of calcium hardness?

Given: [EDTA] = 0.0080 M, Sample Volume = 50.00 mL, Titrant Volume = 10.68 mL of EDTA.

Find: [Ca hardness] = ?

$$[\text{Ca Hardness}] = \frac{(\text{molarity of EDTA})(\text{mL of EDTA added})}{\text{mL of sample volume}}$$

$$[\text{Ca Hardness}] = \frac{(0.0080M)(10.68)}{50.00} = 0.0017M$$

Now to convert to mg/L as $CaCO_3$, simply multiply by 50,000. Why 50,000? The answer is because the molecular weight of calcium carbonate is 100 and there are 2 equivalents of charge, so 100 ÷ 2 = 50, and there are 1000 mg/g, so 50 × 1000 = 50,000.

$$\text{Ca Hardness} = 0.0017M \times 50,000 = 85 mg/L \text{ as } CaCO_3$$

Pre-Lab Questions

1. What does hardness mean?

2. Look up typical values for hardness for natural water samples in your area (provide references).

3. Other than calcium and magnesium, what other cations might contribute to the hardness in your samples?

4. What is meant by carbonate vs. noncarbonate hardness?

5. If one sample has Ca^{2+} = 175 mg/L and Mg^{2+} = 22.5 mg/L, calculate the total hardness in mg/L as $CaCO_3$.

Procedure

1. **Obtain sample.** Each group should provide **two** environmental samples. Collect **2 liters** of each sample. One sample should be rain water. If you cannot collect a rain water sample, try one of these: river water or canal water or bottled water. The second sample should be sea water. A third QA/QC sample will be provided.

2. **Determine the expected total hardness.** Take one test strip being careful not to touch the color squares with your fingers or gloves, and dip the strip into the sample until the color no longer changes. Pull out the strip, and compare the color to the color standards on the package label (Figure 41). This is the *expected total hardness*.

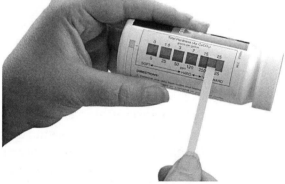

Figure 41. Hardness test strip
Courtesy of Florida Atlantic University

3. **Select an appropriate sample volume and titration cartridge** corresponding to the *expected alkalinity* from Table 11.

Table 11. Sample volumes for hardness

Range (mg/L as CaCO₃)	Sample Volume (mL)	Titration Cartridge (M EDTA)	Digit Multiplier
10 – 40	100	0.0800	0.1
40 – 160	25	0.0800	0.4
100 – 400	100	0.800	1.0
200 – 800	50	0.800	2.0
500 – 2000	20	0.800	5.0
1000 – 4000	10	0.800	10.0

Optional: If the expected sample concentration is not known, start with a smaller sample volume and determine its approximate concentration. Retest with the appropriate sample size.

4. **Prepare the digital titrator.** Now, slide the appropriate cartridge into the titrator receptacle and lock in position with a slight turn. Remove the polyethylene cap and insert a clean delivery tube into the end of the cartridge until tight. To start titrant flowing, hold the tip of the cartridge up. Advance the plunger release button to engage the piston with the cartridge (push the button in and toward the cartridge). *Do not expel solution when pushing the piston toward the cartridge.* Turn the delivery knob until air is expelled and several drops of solution flow from the tip. Then use the counter reset knob to turn the digital counter back to zero and wipe the tip.

5. **Prepare the sample.** Use a graduated cylinder or pipet to measure the appropriate sample volume indicated from Table 11. Transfer the sample into a clean 300-mL beaker. If necessary, dilute with deionized water and make a note to adjust your final result using the appropriate digit multiplier. Make sure your first sample is the QA/QC check standard provided by your instructor.

6. **Evaluate calcium hardness.** Perform titration as follows.

7. **Add 2.0 mL of 8.0 N Potassium Hydroxide Standard Solution.** Swirl to mix.

8. **Add the contents on one CalVer® 2 Calcium Indicator Powder Pillow** Swirl to mix.

9. **If a pink color appears, add titrant (EDTA) dropwise, to a blue endpoint.** Make sure to place the delivery tube tip into the solution and swirl the flask while titrating. Record the number of digits on the titrator (or better yet, record both initial and final readings) in a table like the one that follows.

Sample Identification	Sample Vol. (mL)	Cartridge (M)	Initial Digits	Final Digits	Multiplier

10. **Compute the Calcium Hardness.** Calculate the calcium hardness value by multiplying the digits on the digital titrator (Final − Initial) by the digit multiplier. Report all values in mg/L as $CaCO_3$. **Do not reset counter to zero!**

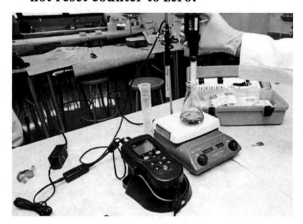

Courtesy of Florida Atlantic University

11. **Evaluate total hardness.** After completing the calcium hardness titration, add 1.0 mL of 5.25 N Sulfuric Acid Standard until the color changes from blue to purple, then to blue, and finally to red. Swirl to mix and to ensure that all precipitated magnesium hydroxide has re-dissolved.

12. **Add 2.0 mL of Hardness 1 Buffer Solution.** Swirl to mix.

13. **Add the contents of one ManVer® 2 Hardness Indicator Powder Pillow or 4 drops of Hardness 2 Test Solution.** Swirl to mix.

14. **Begin counting from the previous endpoint, and titrate from red to blue.** The endpoint looks purple then goes to a true light blue color (record your readings in a table). Calculate the total hardness value by multiplying the digits on the digital titrator by the digit multiplier. Report all values in mg/L as CaCO₃.

15. **Compute the magnesium hardness by difference.** The first titration yields calcium hardness. The second titration gives the total hardness. Therefore, the difference will be the magnesium hardness.

16. **Take 100 mL of your sample and add 1 drop of diluted (10%) dishwashing detergent solution (Liquinox®).** Stir for 30 seconds, let stand for 30 seconds and transfer about 10 mL to a 40 mL I-Chem vial. Shake well and after letting stand for 30 seconds, measure the height of the bubble rise from the air-water interface for all of your samples using a ruler and record these values in a table, like the one that follows:

Sample	Ca Hardness (mg/L as CaCO₃)	Mg Hardness (mg/L as CaCO₃)	Total Hardness (mg/L as CaCO₃)	Bubble Height (mm)
Unknown				
Rain				
1:1 Rain				
Seawater				
1:1 Seawater				
Dilution Water				

Discussion Questions

1. Plot a bar chart showing your samples in increasing order of total hardness.

2. What problems and consumer complaints related to hardness would you expect, if any, if these samples were in fact tap water?

3. Plot a graph showing the height of soap suds vs. total hardness for all of your samples. Explain why you observe more soap suds in one sample versus all the others.

4. Which one of your samples is better for cleaning with? Explain.

5. If you titrate 500.0 mL of sample with 0.100 M EDTA and the titration requires 2.575 mL to reach the total hardness endpoint, what is the calcium hardness in mg/L as CaCO₃?

Proper Safety and Disposal

Dispose of all of your samples in the appropriate waste container.

References

- 29 CFR 1910.1200: Hazard Communication
- 29 CFR 1910.1450: Chemical Hygiene, Occupational Exposure to Hazardous Chemicals in Laboratories
- 29 CFR 1920.132: Personal Protective Equipment
- 29 CFR Subpart Z: Toxic and Hazardous Substances
- 40 CFR 261: Environmental Protection Agency, Resource Conservation and Recovery Act
- Beer, D. F., & McMurrey, D. A. (2005). *A Guide to Writing as an Engineer.* New York: Wiley.
- Bridgeford, T., Kitalong, K. S., & Selfe, R. (2004). Innovative approaches to teaching technical communication.
- Ebel, H. F., Bliefert, C., & Russey, W. E. (2004). *The art of scientific writing: from student reports to professional publications in chemistry and related fields.* John Wiley & Sons.
- *Fundamentals of General, Organic and Biological Chemistry, Media Update Edition, 4/E.* McMurry, J. and Castellion, M.E., Prentice Hall, 2006
- Hargis, G., Carey, M., Hernandez, A. K., Hughes, P., Longo, D., Rouiller, S., & Wilde, E. (2004). *Developing quality technical information: A handbook for writers and editors.* Pearson Education.
- Houp, K. W., Pearsall, T. E., Tebeaux, E., & Dragga, S. (1998). *Reporting technical information.* Boston: Allyn and Bacon.
- Keeling, C. D., Piper, S. C., Bacastow, R. B., Wahlen, M., Whorf, T. P., Heimann, M., & Meijer, H. A. (2001). Exchanges of atmospheric CO_2 and $^{13}CO_2$ with the terrestrial biosphere and oceans from 1978 to 2000. I. Global aspects. *Scripps Institution of Oceanography.*
- Langelier, W.F. (1936). The analytical control of anti-corrosion water treatment. *Journal of the American Water Works Association,* 28: 1500.
- Mann, M. E., Bradley, R. S., & Hughes, M. K. (1998). Global-scale temperature patterns and climate forcing over the past six centuries. *Nature, 392*(6678), 779-787.
- Masters, G. M., & Ela, W. (2008). *Introduction to environmental engineering and science* (Vol. 3). Englewood Cliffs, NJ: Prentice Hall.
- Metcalf, E. E., & Eddy, H. (2003). Wastewater engineer treatment disposal, reuse. New York: McGraw Hill.
- Nagle, J. G. (1996). *Handbook for preparing engineering documents: from concept to completion.* John Wiley & Sons.
- *Prudent Practices in the Laboratory Handling and Disposal of Chemicals,* National Research Council, 1995
- *Safety in Academic Chemistry Laboratories,* American Chemical Society, 1990
- Saltzman, J. (1993). *If you can talk, you can write.* Warner Books.
- Silyn-Roberts, H. (2012). *Writing for science and engineering: Papers, presentations and reports.* Newnes.
- Skoog, D. A., West, D. M., Holler, F. J., & Crouch, S. (2013). *Fundamentals of analytical chemistry.* Nelson Education.
- Socha, T. (2001). *A Technical Guide for Performing and Writing Phase I Environmental Site Assessments.* iUniverse.
- Tchobanoglous, G. and Schroeder, E.D. (1985). Water Quality. Addison-Wesley, Reading, MA.
- Thomas, H. A. (1937). The" slope" method of evaluating the constants of the first-stage biochemical oxygen-demand curve. *Sewage Works Journal,* 425-430.
- Winsor, D. A. (2012). *Writing power: Communication in an engineering center.* SUNY Press.

		Excellent	Good	Fair	Poor	Not Acceptable
☑	**Title Page**	Contains informative title, names, date, course number and course name, teacher, experiment number, and keywords	Missing one item, except title or names	Missing two items, except title or names	Missing more than two items, or title or names, or title merely restates the handout	No title page
•	**Descriptive**	Titles should be informative, have less than 15 words, should not restate the title of the handout, and 5 – 6 keywords should be listed at the bottom of the page				
•	**Includes date and names**	Only those names on the title page will get credit for participating Must include the date of the experiment and the date of the submittal (see Title page above)				
☑	**Table of Contents**	Contains a list of headings for all essential elements with page numbers	Formatting error or one missing element	Formatting error and one missing element	Formatting error and more than one missing element	No table of contents included
☑	**List of Figures**	List of all figures (separately) with page numbers	Combined table or one missing figure	Combined table and/or missing few figures	Combined table and/or missing several figures	No lists included
☑	**List of Tables**	List of all tables (separately) with page numbers	Combined table or one missing table	Combined table and/or missing few tables	Combined table and/or missing several tables	No lists included
☑	**Executive Summary**	Stand alone (<1 page and 3 paragraphs) abstract with all essential elements	Too long or too short or missing one of the essential elements	Too long or too short and missing one of the essential elements	Too long or too short and missing more than one of the essential elements	No summary included
•	**¶1 – Intro**	Brief synopsis of the topic, purpose of experiment, all objectives clearly stated				
•	**¶2 – Methods**	Brief description of how each of the objectives will be tested				
•	**¶3 – Major Findings**	Brief summary of major findings and conclusions including **specific data results** that directly address each of the stated objectives				
☑	**Introduction**	Defines scope, refers to pertinent literature with proper references, ends with objectives	Gives a correct purpose with some framework	Declares a purpose that is correct, weak discussion of the topic	Purpose is incorrect, discussion merely copied from handout	No introduction included
•	**Define scope**	Provides a brief synopsis of the essential background to understand the topic, clearly explains the main concepts, educates by providing context, reviews the pertinent literature, and addresses the purpose of the experiment				
•	**Refers to literature**	A minimum of 5 references are required, of these 3 must be peer-reviewed (book or journal publication) only 2 may be from websites, always paraphrase (avoid direct quotes) in the text, **cite sources (author year)**				
•	**Ends with objectives**	States the hypothesis clearly (mentions the anticipated outcomes) and finishes with the objectives of the experiment and how the experiment will address those objectives				

		Excellent	Good	Fair	Poor	Not Acceptable
☑	**Methodology**	Detailed step-by-step description of procedures and QA/QC, labeled diagrams or photos of equipment and materials used	Step-by-step description that misses not more than one key detail, diagrams/drawings are not labeled	Step-by-step description that misses not more than two key details; equipment only mentioned	Description lacks more than two key details; no mention of equipment used to carry out the experiment	No methodology or merely copied from handout
•	**Experimental setup**	Sketch the setup (include details, dimensions, or photos), discuss the procedure in the order in which the steps were performed (do not copy the handout because things change sometimes), describe how experimental groups differ from the controls				
•	**Measurements and Analytical Methods**	Describes what was recorded and how the data were collected. Describes statistics used, describes quality assurance/quality control (QA/QC) procedures				
☑	**Results and Discussion**	Explanation of results with error discussion and hypothesis testing, Calculations are correct and clearly shown, Figures display data correctly, all variables labeled, units are correct and consistent	Explanation of results with error discussion and hypothesis testing, Calculations contain few errors in units or math, Figures correct, variables unlabeled, and units are correct and consistent	Explanation of results with error discussion and hypothesis testing, Calculations contain errors in units or math, Figures correct. No labels or legend, and units are correct and consistent	Math not shown. Figures and tables display data incorrectly, and units are correct and consistent	No results presented, no discussion of trends, no tables/figures included, or weak data analysis
•	**Data and Interpretation**	Describes how data were collected and analyzed. Results are explained not merely presented. Discusses each figure, table, sketch, or photo included in the text and refer to each one. States observations. **Calculations must be correctly done**. At least one sample calculation must be included in the text. Discusses observations and trends. Did the data correspond to what you expected? Why or why not? Did the data correspond to what other people found? (use citations here). If no, why do you think it happened? Present a new hypothesis. If yes, are there other factors that may have caused this outcome?				
•	**Tables and figures**	Presents observations and results in graphical, tabular, or sketch format, follows all rules for tables/figures format, includes proper units and labels, raw data goes in the appendix, tables/figures are numbered independently, all mentioned in the text				
•	**Analysis**	Discusses correlations, trends, errors, or outliers observed Justifies statements with results from statistical analysis Clearly states if the hypothesis was supported or not by the observed data				
•	**Discussion questions**	Each worth one point apiece, answered after the "Results and Discussion" section, but before the "Conclusions" section				

		Excellent	Good	Fair	Poor	Not Acceptable
☑	**Conclusions**	Summary of major findings, uses data to support results of hypothesis testing, discusses errors, addresses broader impacts	Missing one item, except summary of major findings with data to support, addresses broader impacts	Missing two items, except summary of major findings with data to support	Missing more than two items	Poorly written conclusion
•	**Hypothesis testing**	Restates the hypothesis, supports or refutes it, and explains the role of the experiment and the data collected in making your decision				
•	**Supporting evidence**	States the major findings and uses collected data as evidence to support the findings Summary is logically arrived to from collected data and prior knowledge				
•	**Error discussion**	Identifies sources of error and explains effect on results Provides thoughtful questions and suggestions to improve quality of results				
•	**Broader impacts**	The experiment is made meaningful by discussion of its scientific or societal implications. Did the findings stimulate other questions for further research?				

Laboratory Report Scoring Sheet

Title Page

Descriptive	Includes Date/Names
Descriptive	Names
Less than 15 words	Course and teacher
Not restated handout title	Activity number and date
5 – 6 keywords	Date of submittal

(Front Matter)

Table of Contents	List of Figures	List of Tables
Subheadings for all essential elements	List of all figures (separately)	List of all tables (separately)
Page numbers	Page numbers	Page numbers

Executive Summary

¶1 – Intro	¶2 – Methods	¶3 – Major Findings
Brief synopsis	Description of how each objective will be tested	Brief summary of major findings
Purpose of experiment		Specific data results that address each objective
Objectives clearly stated		

Introduction

Defines Scope	Refers to Literature	Ends with Objectives
Brief background	Reviews pertinent literature	Addresses the purpose
Explains main concepts	5+ references required	States hypothesis
Educates by providing context	3 must be peer-reviewed	Mentions anticipated outcomes
Avoids any obvious gaps	Only 2 from websites	Finishes with all objectives
Understanding of the topic	Cites sources (author year)	

Methodology

Experimental Setup	Measurements and Analytical Method
Sketch the setup (details, photos)	What was recorded
Complete accurate procedure	How was data collected
Not copied from manual	Describes statistics used
Paragraph form, past tense, passive voice	Describes QA/QC

Results and Discussion

Data and Interpretation	Tables and Figures	Analysis
How data was collected/analyzed	Format with proper units/labels	Explains results
State observations	Descriptive captions	Correlation/trends
Sample calcs shown	Source credit given	Errors/outliers
Calcs correct	Referred to in the text	Evidence/statistics
Expected/unexpected	Effective presentation	Hypothesis justified or not

Conclusions

Hypothesis Testing and Supporting Evidence	Error Discussion	Broader Impacts
Restates the hypothesis	Identifies errors	Summary is logical
Reports major findings	Explains impact on results	Science/societal implications
Uses data to support statements	Thoughtful suggestions to improve the results	Thoughtful questions
Explains why results support or contradict expected outcomes		Further research

(Back Matter)

References	Appendix
5+ references required	Correctly arranged
3 must be peer-reviewed	Correctly labeled
Only 2 from websites	1st page of citation
Cites sources (author year)	
ASCE journal format	

Writing Style

Grammar and Syntax	Overall Impression
Spell/grammar check	Has strengths
Tone	Shows conceptual understanding
Sentence structure	Cohesive, clear, concise
Rhetorical structure	Responds to the purpose
Sentence-level patterns of error	

Writing Style (formatting)

1-inch margins
1.5-spaced
Times or Arial 11 pt
Block justification

Discussion Questions

1	2	3	4	5	6	7	8	9	10	11	12	13	14	15	TOTAL

Designed by Daniel E. Meeroff, Ph.D. ©2006

Version 2013